Principia Philosophiæ Scientiæ

(The Principles of the Philosophy of Science)

I0465502

Nader Haddad

"Doubt is the beginning, not the end, of wisdom." — Socrates

Published by Nader Haddad Paris, 2024 Distributed by Amazon KDP

Copyright Page

Principia Philosophiæ Scientiæ
(The Principles of the Philosophy of Science)

ISBN: 9798300713454

Published by Nader Haddad via Amazon KDP

This book is a work of nonfiction. While every effort has been made to ensure the accuracy of the information contained herein, the author assumes no responsibility for errors or omissions or for any consequences arising from the use of the information in this book.

Cover Design by: Canva ·
Interior Formatting by: Canva

First Edition

Printed in the United States of America ,

 United Kingdom , Oxford ,

France

Acknowledgments

This book is dedicated to the memory of my father, who, though not a scientist, gave me the most powerful tool of discovery: the courage to ask questions. His wisdom and curiosity continue to guide me, even years after his passing.

To my family, your unwavering love and support have been the foundation of my journey. You remind me every day of the importance of understanding, compassion, and the pursuit of meaning.

I also owe a deep sense of gratitude to the life and legacy of John Nash. His extraordinary contributions to mathematics and his resilience in facing the challenges of mental illness have profoundly inspired me. His story is a testament to the triumph of the human spirit and the boundless capacity of the mind, even in adversity.

This book is a reflection of these influences—a journey into the questions that shape our understanding of the world and ourselves.

The Author

A Letter from the Author

To those who recognized the shadow of my sorrow and did not turn away, who instead chose to see me, truly see me, in my vulnerability and solitude—to you, I owe a debt of love that words cannot fully repay. You did not dismiss my melancholy as weakness, nor did you leave me to wander alone through the labyrinth of my own despair. Instead, you sought to unravel its mysteries with care, offering your presence as a guiding light.

To those who placed me above others, even when I faltered, even when I failed you, your unwavering grace humbles me. You chose to see the best in me when I could not see it in myself, and for that, I owe you my gratitude, my devotion, and my love. To those who embraced me with all my flaws, who defended my honor despite my shortcomings, and who stood by me when others might have walked away—you are the noble souls who remind me that true strength lies in kindness.

To those who entered my life not as tempests but as gentle breezes, bringing comfort rather than chaos, to you I owe a peace that soothes the storms within me. Your presence has been a balm, quieting the tumult of uncertainty and offering instead a steadying hand. To those who chose gentleness in their words and actions, whose kindness never turned sharp or cruel even when it could have—your compassion has been a gift, one that I carry with me always.

To those who sought me out in my moments of solitude, when psychological distress rendered the world a harsh and unforgiving place, you gave me the most precious gift of all: your time, your understanding, and your refusal to let me face suffering alone. To those in whom I placed my trust, and who carried that trust as a sacred bond, unbroken and unwavering, you are the faithful guardians of my soul. To you, I owe love, in its purest and most enduring form.

This work, **Principia Philosophiæ Scientiæ**, is not merely a collection of ideas—it is a fragment of my very being, a piece of my soul rendered in words. It is both a tribute and a testament, born from the dual forces of love and doubt, shaped by the interplay of light and shadow in my own life. To write this book has been to place myself upon the page, to open my heart to those who will read it, and to offer the lessons I have gleaned through inquiry, struggle, and perseverance.

As I wrote, I carried with me the weight of gratitude for those who walked this path alongside me, whether they realized it or not. To who stayed when I faltered, who mended what was broken, and who stood as pillars of strength when mine crumbled—you are the quiet heroes of this journey. Your care gave me the courage to press forward, to navigate the complexities of doubt and discovery, and to find meaning even in uncertainty.

This book is for you, my companions in thought and spirit. It is an offering of thanks and a reflection of the love you have given me. Within these pages are not just the musings of a thinker or the discoveries of a seeker, but the echoes of the kindness, patience, and resilience you have shown me.

To all those who chose compassion over judgment, understanding over rejection, and love over indifference, I dedicate this work. It is a fragment of my soul, woven together with the threads of your care and the light of your faith in me. It is my hope that these pages will not only illuminate the questions of science and philosophy but also serve as a reminder of the power of connection, the enduring strength of love, and the transformative potential of doubt.

With all the gratitude my heart can hold,
Nader Haddad

Preface

What drives humanity to seek answers? Why do we gaze at the stars, question the nature of reality, or challenge the limits of our understanding? These questions have lingered in the minds of great thinkers for millennia, and they remain as urgent today as they were in the marketplaces of ancient Athens or the laboratories of modern physics.

This book is born from that same curiosity—the need to know, to doubt, and to wonder. It is a journey through the intertwined histories of science and philosophy, disciplines that have shaped the course of human progress. While science strives to uncover the mechanics of the universe, philosophy seeks to interpret its meaning. Together, they reveal not only the vastness of the cosmos but also the depths of the human mind.

But this journey is also deeply personal. It is inspired by the wisdom of my father, who taught me that asking the right question is often more important than finding the answer. He showed me that curiosity is the essence of discovery, a lesson that has shaped my life and this book.

I have also drawn inspiration from the story of John Nash, whose brilliant mind reshaped mathematics while battling the challenges of mental illness. His life taught me that resilience and creativity often emerge from the very struggles we face, reminding us of the incredible complexity of human existence.

In an age defined by rapid scientific advancement and societal challenges, this book invites you to reflect on the principles that underlie our understanding of the world. It is an exploration of truth, doubt, and the extraordinary power of the human mind to see beyond illusion and question the status quo.

Whether you are a lover of science, a seeker of philosophical insight, or simply someone curious about the nature of existence, I hope this book will inspire you to think more deeply, question more boldly, and embrace the uncertainties that make discovery possible.

Welcome to *Principia Philosophiæ Scientiæ.*

Table of Contents

- Examples: Inattentional blindness, anchoring bias, optimism bias.

9. *Scientia, Philosophia, et Existentia* (Science, Philosophy, and Existence)
   o The search for meaning in an indifferent universe.
   o Examples: Split-brain experiments, Einstein's relativity.
10. *De Dubio et Progressu* (On Doubt and Progress)
    o How doubt drives innovation.
    o Examples: Socrates' method, Einstein's "blunder," Feynman's principles.

---

11. **Conclusion: Ad Principia Novae Cogitationis**
    o Synthesis of lessons from science and philosophy.
    o A call to action for critical thinking and intellectual humility.

Introduction: De Origine Scientiae et Philosophiae

(On the Origins of Science and Philosophy)

Humanity has always sought to understand the universe. From the first glimmers of consciousness, we gazed at the stars, listened to the rustling leaves, and wondered: *What is all this? And why does it exist?* This primal curiosity gave birth to two intertwined pursuits: philosophy, the love of wisdom, and science, the systematic exploration of the natural world.

Though they often appear as separate disciplines today, science and philosophy began as one. In the bustling marketplaces of ancient Athens, philosophers like Socrates asked questions not just about ethics and politics but about the nature of reality itself. In the temples of Egypt and Mesopotamia, priest-astronomers tracked the heavens, blending celestial observations with metaphysical speculation. Both disciplines sought the same goal: to unveil truth, however elusive or inconvenient it might be.

But truth is not easily grasped. What we see is not always true, and what we think is not always accurate. A stick submerged in water appears bent; the sun seems to rise and set around us. Our minds, shaped by evolution to survive rather than comprehend, are prone to error. This book is a journey into the heart of this paradox, exploring how science and philosophy have grappled with the fundamental unreliability of perception and reason.

The chapters that follow are divided into three parts:

1. **The Foundations of Science:** How the human desire to understand led to systematic methods of inquiry.
2. **The Philosophy of Science:** The theoretical underpinnings that challenge and refine our quest for knowledge.
3. **Science and Humanity:** How cognitive biases and philosophical insights shape our understanding of existence.

Let us begin with a simple proposition: *Not everything we see is true, and not everything we think is true.* From this premise, we will explore the enduring partnership between science and philosophy and their shared mission to reveal the deeper truths of reality.

Part I: Fundamenta Scientiae

(The Foundations of Science)

Science is born of curiosity and doubt. It begins with the question *Why?* and moves forward through observation, experimentation, and the search for evidence. But the journey of science is rarely straightforward. It is shaped by the limits of human perception, the errors of reasoning, and the illusions that often mask the truth.

In this part, we explore the foundations of science: how questioning became a method, how perception both guides and deceives us, and how errors and failures have shaped the greatest discoveries.

Chapters in Part I:

- **Chapter 1: De Natura Scientiae** *(On the Nature of Science)*
- **Chapter 2: Scientia et Illusio** *(Science and Illusion)*
- **Chapter 3: De Ratione et Falsitate** *(On Reason and Falsehood)*

Begin Chapter 1: De Natura Scientiae

Chapter 1: De Natura Scientiae

(On the Nature of Science)

Science is an act of rebellion. It rebels against complacency, against the assumption that what we observe or believe is necessarily true. To question, to test, to verify—these are the foundations of the scientific method, a framework that emerged not from a single moment in history but from centuries of trial, error, and reflection.

In ancient Greece, Thales of Miletus proposed that water was the fundamental substance of the universe, not because he had conclusive evidence, but because he dared to ask a different question: *What if the world is not governed by the whims of gods but by principles we can understand?* This marked the birth of natural philosophy, the precursor to modern science.

Yet even as early thinkers sought to explain the world rationally, they encountered a persistent challenge: the unreliability of human perception. Consider the stars. To the naked eye, they appear fixed in the sky, circling above us each night. For centuries, this perception reinforced the belief in a geocentric universe, where Earth stood immovable at the center of creation. It was a model that seemed logical, even self-evident—until it wasn't.

The Transformative Power of Doubt

The Copernican Revolution exemplifies the audacity of questioning appearances. Nicolaus Copernicus, a Polish astronomer and cleric, proposed a heliocentric model of the solar system in the 16th century. His idea was not immediately embraced; after all, if Earth moved, why didn't we feel it? Why didn't objects fly off its surface? And yet, through meticulous observation and reasoning, Copernicus and those who followed—Galileo Galilei, Johannes Kepler, and Isaac Newton—dismantled the geocentric illusion, revealing a universe far more complex and beautiful than anyone had imagined.

Scientific Progress Through Mistakes

Even the greatest scientific minds, such as Albert Einstein, were not immune to error. Einstein famously introduced the cosmological constant to force his equations to describe a static universe. Later, when Edwin Hubble discovered the universe was expanding, Einstein called this his "biggest blunder." And yet, this very concept—once dismissed—now plays a critical role in our understanding of dark energy.

Each of these examples highlights the iterative nature of science. Doubt, error, and revision are not signs of weakness but of strength. They are the mechanisms through which science grows closer to the truth.

Chapter 2: Scientia et Illusio

(Science and Illusion)

Introduction: The Deceptive Nature of Perception

"Seeing is believing," or so the saying goes. But human history reveals how often our senses deceive us. From the illusion of a flat Earth to the mind-bending discoveries of quantum mechanics, our perception of reality is rarely aligned with the truth.

Perception is not a passive mirror of the world; it is an active process shaped by the brain, which filters and interprets sensory data. While this filtering enables us to navigate a complex environment efficiently, it also creates illusions—misleading interpretations of reality that can obscure the truth.

This chapter explores how illusions, both sensory and intellectual, have shaped the history of science. By examining examples from the past and the present, we see how the quest to separate illusion from reality drives progress.

The Flat Earth: A Persistent Illusion

In ancient times, the flatness of the Earth seemed self-evident. Sailors saw the horizon stretch endlessly, reinforcing the illusion of a flat plane. Even early scientific theories, such as those of the Babylonians and Egyptians, accepted a flat Earth as fact.

It wasn't until the Greek philosopher Pythagoras, and later Eratosthenes, that the Earth's roundness was established. Eratosthenes' experiment involved measuring the angles of shadows in two cities at the same time. By comparing these angles, he calculated the Earth's circumference with remarkable accuracy.

This breakthrough illustrates how careful observation and reasoning can pierce the veil of sensory illusion, revealing a deeper truth about the world.

Optical Illusions and the Brain

Even in everyday life, optical illusions reveal the limits of perception.

Example: The Müller-Lyer Illusion
In this classic illusion, two horizontal lines of equal length appear unequal because of the orientation of arrow-like tips at their ends. This occurs because the brain interprets the lines in the context of depth cues, distorting our perception of their true length.

Such illusions highlight a fundamental truth: what we see is not always what is. Our brains actively construct reality based on patterns, context, and prior experience, making us susceptible to errors.

Galileo's Telescope: Shattering Assumptions

The power of science lies in its ability to extend perception beyond the limits of the senses. Galileo Galilei's telescope is a perfect example of this.

When Galileo turned his telescope to the heavens in 1610, he made observations that challenged the long-held belief in a geocentric universe. He discovered craters on the Moon, the phases of Venus, and four moons orbiting Jupiter—all evidence that contradicted the Aristotelian view of perfect, unchanging celestial spheres.

These discoveries were met with resistance, as they threatened both scientific orthodoxy and religious dogma. Yet Galileo's work demonstrated that tools and methods could reveal truths hidden from human perception, expanding our understanding of the cosmos.

Quantum Mechanics: The Limits of Intuition

In the 20th century, quantum mechanics revealed a reality far stranger than anything imagined by classical physics.

Example: The Double-Slit Experiment

When particles such as electrons are fired at a barrier with two slits, they create an interference pattern on the other side—behavior characteristic of waves. But when scientists place detectors at the slits to observe which path the electrons take, the interference pattern disappears, and the particles behave like solid objects.

This experiment defies intuition, challenging our understanding of particles and waves. It suggests that observation itself affects the outcome, highlighting the limits of human perception and the need for theoretical models to grasp the deeper nature of reality.

Mirages and Atmospheric Illusions

Even the natural world creates illusions that challenge our understanding. Mirages, for instance, are caused by the refraction of light in layers of air with differing temperatures. A common example is the illusion of water in the desert, where light bends to create the appearance of a shimmering pool.

For centuries, such phenomena were explained through myths and superstitions. It was only through the scientific study of optics that these illusions were demystified, revealing the interplay between light, air, and heat.

Illusions in Modern Science: Dark Matter and Dark Energy

In the modern era, science continues to grapple with phenomena that challenge perception.

Dark Matter: Observations of galaxies reveal that visible matter accounts for only a small fraction of the gravitational effects observed. The remaining mass—known as dark matter—is invisible, detectable only through its gravitational influence.

Dark Energy: Similarly, the accelerated expansion of the universe points to the existence of dark energy, a force that remains poorly understood. These discoveries remind us that much of reality lies beyond the reach of our senses, requiring sophisticated tools and theories to uncover.

Conclusion: The Role of Science in Unmasking Illusion

Illusions, whether caused by sensory limitations or cognitive biases, are barriers to understanding. Yet they are also opportunities for discovery. By questioning appearances and developing tools to extend perception, science continually pushes the boundaries of knowledge.

As we move to the next chapter, we will explore the errors of human reasoning and how they influence our quest for truth. From logical fallacies to cognitive traps, we will see how philosophy and science together navigate the pitfalls of thought.

Chapter 3: De Ratione et Falsitate

(On Reason and Falsehood)

Introduction: The Fallibility of Reason

Reason is one of humanity's greatest tools for understanding the world. It allows us to create logical systems, make predictions, and solve problems. Yet, reason is far from infallible. Throughout history, even the most rigorous logical frameworks have led to errors— sometimes because of flawed assumptions, sometimes because of the inherent limitations of human cognition.

In this chapter, we explore how reasoning itself can go astray. From the persistence of geocentric models to the biases that cloud modern decision-making, we will examine how philosophy and science work together to identify and correct these errors.

The Persistence of Falsehoods: The Geocentric Model

For centuries, the geocentric model of the universe reigned supreme. Rooted in the ideas of Aristotle and Ptolemy, it placed Earth at the center of creation, with the Sun, Moon, and stars orbiting around it. This system, while incorrect, was elegant and logical—within its own framework.

Flawed Assumptions:
The geocentric model relied on two key assumptions:

1. That Earth was stationary, as no motion was felt.
2. That celestial bodies moved in perfect circles, reflecting divine harmony.

To account for discrepancies, such as the retrograde motion of planets, Ptolemaic astronomers introduced increasingly complex epicycles—small circular orbits superimposed on larger ones. These adjustments preserved the model's internal consistency but moved further from the truth.

The Role of Doubt:
It took Copernicus, and later Galileo, to challenge these assumptions and propose a heliocentric model. Even then, resistance persisted. This resistance illustrates how logical reasoning, when built on false premises, can lead to elaborate but ultimately incorrect systems.

Logical Fallacies: Traps of Thought

The human mind is prone to errors in reasoning, many of which have been cataloged as logical fallacies. These fallacies distort arguments and hinder the pursuit of truth.

Example: The Appeal to Authority
The geocentric model's longevity was partly due to the authority of Aristotle, whose ideas were held in almost religious reverence. This appeal to authority discouraged critical questioning, delaying progress for centuries.

Example: Post Hoc Ergo Propter Hoc
This fallacy, meaning "after this, therefore because of this," assumes causation from correlation. An example from history is the belief that diseases were caused by "miasma" or bad air because outbreaks often occurred in foul-smelling areas. It wasn't until the germ theory of disease was developed that the true causes were understood.

Cognitive Biases: The Mind's Blind Spots

Modern psychology has identified numerous cognitive biases that distort reasoning. Unlike fallacies, which occur in arguments, biases are subconscious tendencies that influence how we process information.

Confirmation Bias:
This bias leads people to seek out information that supports their existing beliefs while ignoring evidence that contradicts them.

Example in Science:
The initial rejection of Alfred Wegener's theory of continental drift is a classic example. Despite compelling evidence, geologists dismissed his ideas because they conflicted with the prevailing view of an immobile Earth. Only decades later, with the discovery of plate tectonics, was Wegener vindicated.

The Dunning-Kruger Effect:
This bias causes individuals with limited knowledge in a domain to overestimate their competence, while experts tend to underestimate theirs.

Example:
In debates on climate change or vaccination, non-experts often feel confident dismissing the overwhelming consensus of scientific evidence, believing their limited understanding to be sufficient.

Errors in Philosophical Reasoning: David Hume on Causation

Scottish philosopher David Hume posed one of the most profound challenges to human reasoning: the problem of causation. Hume argued that we never directly observe causation itself—only sequences of events. For example, when we see one billiard ball strike another and the second ball move, we infer that the first caused the second to move. But this inference is based on habit, not certainty.

Philosophical Implication:
Hume's skepticism about causation laid the groundwork for modern scientific inquiry, which relies on controlled experiments to establish causal relationships. It also influenced later thinkers like Immanuel Kant, who sought to reconcile the limits of human reasoning with the search for universal truths.

Modern Errors: Overconfidence in Predictive Models

Even in the age of supercomputers and big data, reasoning errors persist. Predictive models, while powerful, are often limited by the assumptions and data they rely on.

Example: The 2008 Financial Crisis
In the lead-up to the crisis, financial institutions relied on models that underestimated the likelihood of a systemic collapse. These models assumed that housing prices would continue to rise and ignored warning signs of instability. The overconfidence in these models led to one of the most severe economic downturns in modern history.

Lesson:
This example underscores the importance of questioning assumptions, especially in complex systems. Reasoning, no matter how sophisticated, must remain open to revision in light of new evidence.

The Role of Philosophy in Correcting Errors

Philosophy complements science by providing tools to analyze and refine reasoning. The principles of logic, skepticism, and critical thinking have been essential in identifying and addressing the errors that arise in both thought and practice.

Example:
Karl Popper's concept of falsifiability revolutionized the philosophy of science by emphasizing that a theory must be testable and capable of being proven wrong. This principle

has become a cornerstone of scientific inquiry, ensuring that reasoning is continually subjected to scrutiny.

Conclusion: Embracing Error as a Path to Truth

Reason is a powerful tool, but it is not perfect. Its fallibility is not a weakness but a reminder of the importance of humility and doubt in the pursuit of knowledge. By recognizing and addressing errors in reasoning, science and philosophy together move closer to understanding the complexities of reality.

In the next section, we shift our focus from the foundations of science to its philosophical underpinnings. We will explore the nature of truth, the limits of perception, and the enduring tension between what we know and what we can only speculate.

Chapter 4: De Veritate et Perceptione

(On Truth and Perception)

Introduction: What Is Truth?

What is truth? At first glance, the answer seems simple: truth is what corresponds to reality. But this definition quickly unravels under scrutiny. How do we define reality? Is it what we perceive with our senses? What happens when our senses deceive us?

Philosophers have long grappled with these questions. Plato distinguished between the world of appearances and the world of Forms, arguing that true reality lies in unchanging, perfect concepts accessible only through reason. Centuries later, Kant argued that we cannot know the world as it is (*noumena*); we can only know it as it appears to us (*phenomena*), filtered through the categories of human thought.

Science, in its pursuit of objective truth, attempts to separate perception from reality. Yet, it often confronts the same challenges that philosophy has explored for millennia.

Plato's Allegory of the Cave

In his famous allegory, Plato describes prisoners chained in a cave, facing a wall. Behind them, a fire casts shadows on the wall, created by objects they cannot see. To the prisoners, these shadows are reality. But if one prisoner escapes and sees the world outside, they realize the shadows were mere illusions.

Philosophical Implication:
Plato's allegory illustrates the gap between perception and truth. What we perceive may be only a shadow of reality, distorted by the limitations of our senses and understanding.

Scientific Parallel:
Modern science echoes this idea. For centuries, the sky appeared to be a fixed dome with stars embedded in it. Only through observation and theoretical reasoning did we discover that stars are distant suns scattered across an expanding universe.

Truth and Relativity

Albert Einstein's theory of relativity challenged classical notions of absolute truth. According to relativity, measurements of time, space, and even mass depend on the observer's frame of reference. For example, a clock on a fast-moving spaceship will tick more slowly relative to a stationary observer on Earth.

Philosophical Implication:
Relativity shows that truth is not always absolute; it can be contextual, dependent on the conditions under which it is observed.

Everyday Example:
Imagine two people watching a lightning storm from different locations. One sees a bolt strike a tree, while the other sees it strike a rock at the same time. Both observations are true within their own frames of reference, but neither captures the full reality.

Perception as a Filter

Modern neuroscience supports the idea that perception is not a passive reflection of reality but an active process shaped by the brain.

Example: The Blind Spot
The human eye has a blind spot where the optic nerve exits the retina, creating a gap in our visual field. Yet, we never notice this gap because the brain "fills in" the missing information based on surrounding context.

This illustrates a profound truth: perception is not a mirror of reality but a construction, shaped by both sensory input and the brain's interpretive processes.

Objective Truth in Science

Despite the limitations of perception, science strives for objectivity by relying on empirical evidence, repeatability, and rigorous testing.

Example: The Discovery of DNA

The double-helix structure of DNA, uncovered by Watson and Crick in 1953, was not directly observed but inferred from X-ray diffraction patterns. This discovery demonstrates how science extends perception through tools and models, allowing us to uncover truths beyond the limits of the senses.

Philosophical Challenges: Truth vs. Utility

Pragmatist philosophers like William James argued that truth is not an abstract ideal but what works in practice. For example, Newton's laws of motion are not "true" in the sense of being universally valid—they break down at extreme scales—but they are immensely useful for everyday applications.

Scientific Implication:
Science often embraces models that are not strictly true but serve as approximations of reality. This pragmatic approach allows us to navigate a complex world while remaining open to refinement and revision.

Conclusion: The Elusive Nature of Truth

Truth is not a single, unchanging entity. It is shaped by perception, context, and the tools we use to measure and understand it. Science, in its pursuit of truth, must continually question its assumptions, refine its methods, and remain open to new paradigms.

In the next chapter, we will delve deeper into the limits of human knowledge, exploring the concept of uncertainty and the philosophical foundations of epistemology.

Truth in the Age of Simulation: Can We Trust Our Perceptions?

In the modern age, questions about truth and perception take on new dimensions. Advances in technology have allowed us to simulate reality to astonishing degrees. From virtual reality (VR) to artificial intelligence, these technologies blur the line between what is real and what is simulated, challenging the very concept of truth.

Example: The Brain in a Virtual World
Imagine wearing a VR headset that transports you into a hyper-realistic simulation of another world. Your senses—sight, sound, touch—convince you that this world is real, even though you know it is a simulation. Now, extend this thought experiment: How do you know your current reality isn't itself a simulation?

Philosophical Connection:
This question echoes René Descartes' famous thought experiment in *Meditations on First Philosophy*. Descartes imagined an "evil demon" deceiving him into believing a false reality. His radical doubt led him to conclude, *"Cogito, ergo sum"*—"I think, therefore I am"—as the only certainty. Similarly, philosopher Nick Bostrom's "simulation hypothesis" posits that advanced civilizations might create simulated realities indistinguishable from the real world.

Scientific Connection:
Modern neuroscience has shown that the brain is easily fooled into accepting simulations as real. Studies on VR-induced "presence" demonstrate that the brain reacts to virtual environments as if they were genuine. For example, participants in VR experiments often feel fear when standing on virtual ledges, even when they know they are in a safe room.

Implications for Truth:
If perception can be so easily manipulated, how do we distinguish between reality and illusion? This question is not just academic; it has profound implications for how we evaluate evidence in science, ethics, and daily life.

Mathematics: The Search for Universal Truths

While perception may be unreliable, mathematics offers a realm of apparent certainty. Numbers, equations, and logical proofs seem immune to the distortions of the senses. Yet even mathematics grapples with its own challenges to truth.

Example: Gödel's Incompleteness Theorems
In the early 20th century, mathematician Kurt Gödel proved that any sufficiently complex mathematical system contains statements that cannot be proven or disproven within that system. This shattered the dream of finding a complete and consistent set of axioms for mathematics, a dream pursued by thinkers like Bertrand Russell.

Philosophical Implication:
Gödel's work reveals that even in the seemingly pristine world of mathematics, truth is not absolute. There will always be truths that lie beyond formal proof, reminding us of the limits of human understanding.

Scientific Implication:
In physics, mathematical models often predict phenomena long before they are observed. For example, Paul Dirac's equations predicted the existence of antimatter, which was later confirmed experimentally. These cases highlight the dual role of mathematics: as a tool for uncovering truths and as a reminder of the limits of formal systems.

The Role of Cultural and Social Context in Truth

Truth is often influenced by the cultural and social frameworks within which it is pursued. Science, while aiming for objectivity, is not immune to these influences.

Example: The Shifting "Truth" of Disease
In ancient Greece, disease was attributed to imbalances in the four humors: blood, phlegm, black bile, and yellow bile. During the Islamic Golden Age, scholars like Ibn Sina proposed more systematic, evidence-based approaches to medicine. Yet even then, cultural assumptions influenced their conclusions.

It wasn't until the 19th century that germ theory, advanced by Louis Pasteur and Robert Koch, replaced these earlier models. This shift was not merely a scientific breakthrough but a cultural transformation, as societies began to value empirical evidence over tradition.

Modern Implication:
Even today, cultural biases shape scientific inquiry. For example, the underrepresentation of diverse populations in clinical trials can skew medical research, leading to treatments that are less effective for certain groups. Recognizing these biases is essential for advancing a more inclusive and comprehensive pursuit of truth.

Truth and Ethics: The Responsibility of Knowing

The pursuit of truth carries ethical responsibilities. When we uncover truths—about the natural world, human behavior, or technology—we must consider their implications and applications.

Example: The Manhattan Project
The development of nuclear weapons during World War II was a triumph of scientific ingenuity but also a moral quandary. Scientists like J. Robert Oppenheimer grappled with the ethical implications of their work, famously quoting the Bhagavad Gita: *"Now I am become Death, the destroyer of worlds."*

Philosophical Implication:
Truth is not neutral; it exists within a web of moral and social consequences. As philosopher Hannah Arendt argued, the pursuit of knowledge without consideration of its impact can lead to dehumanization and destruction.

Modern Example:
The rise of artificial intelligence poses similar ethical challenges. Truths uncovered in AI research have the potential to revolutionize industries but also to amplify biases, invade privacy, and disrupt social systems. Scientists and philosophers alike must grapple with the ethical dimensions of truth in this rapidly evolving field.

Conclusion: Truth as a Guiding Star

Truth is not a destination but a guiding star. It illuminates our path, even as it recedes into the horizon. From Plato's cave to Gödel's theorems, from VR simulations to the ethics of discovery, the pursuit of truth challenges us to think more deeply, question more boldly, and act more responsibly.

In the next chapter, we will explore the limits of knowledge itself, diving into the philosophical foundations of epistemology and the scientific embrace of uncertainty.

Truth Hidden in the Stars: The Shift from Geocentrism to Heliocentrism

For centuries, the geocentric model of the universe—that the Earth was the center of creation—was considered an unquestionable truth. It was elegant, intuitive, and supported by religious doctrine and the authority of philosophers like Aristotle and Ptolemy. The Sun, Moon, stars, and planets seemed to move around the Earth in predictable patterns. To challenge this view required not only new observations but also a profound shift in how truth was defined.

Nicolaus Copernicus, a 16th-century Polish astronomer, dared to propose a heliocentric model, placing the Sun at the center of the solar system. His revolutionary idea was not born of direct observation but of mathematical simplicity. Copernicus realized that the irregular motions of planets, which required Ptolemy's complex system of epicycles, could be explained more naturally if the Sun, not the Earth, was at the center. However, his model was initially resisted—not just because it defied sensory experience, but because it contradicted deeply held beliefs about humanity's privileged place in the cosmos.

It was only with the work of Galileo Galilei, who used the telescope to observe celestial phenomena, that the Copernican model gained empirical support. Galileo's discoveries—the phases of Venus, the moons of Jupiter, and the imperfections of the Moon's surface—shattered the Aristotelian notion of perfect celestial spheres. Yet Galileo's insistence on the heliocentric model brought him into conflict with the Church, illustrating the social and political barriers to accepting new truths.

This pivotal moment in history demonstrates that truth often lies beyond perception, requiring a synthesis of observation, mathematics, and the courage to challenge authority. It also highlights the philosophical tension between intuitive appearances and the deeper realities uncovered by science.

The Illusion of Solidity: The Atomic Revolution

One of the most profound shifts in our understanding of truth came with the discovery that matter is not solid but composed of tiny, invisible particles. For much of human history, objects appeared to be continuous and unbroken. A rock seemed fundamentally different from water, and both seemed distinct from air. The idea that all matter could be reduced to tiny, indivisible units was first proposed by the ancient Greek philosophers Democritus and Leucippus, but it remained a philosophical abstraction for millennia.

It wasn't until the 19th and 20th centuries that the atomic theory gained empirical support through the work of scientists like John Dalton, J.J. Thomson, and Ernest Rutherford. Rutherford's gold foil experiment, in particular, shattered previous assumptions about the structure of the atom. By firing alpha particles at a thin sheet of gold, Rutherford expected them to pass straight through, as Thomson's "plum pudding model" suggested that atoms were diffuse clouds of positive and negative charge. Instead, some particles were deflected at sharp angles, leading Rutherford to conclude that atoms consisted of a dense nucleus surrounded by mostly empty space.

This discovery fundamentally changed our perception of matter. What we experience as solid—tables, chairs, and even our own bodies—is mostly empty space, held together by invisible forces. This realization not only revolutionized physics but also challenged philosophical notions of substance and identity. If matter is not what it appears to be, how can we trust our senses to reveal the truth?

The Mirage of Time: Einstein's Relativity

Time has long been considered one of the most fundamental aspects of reality. We experience it as a constant, flowing from past to present to future. Ancient philosophers, such as Heraclitus, contemplated the nature of time, famously stating, "You cannot step into the same river twice." For most of history, time was assumed to march uniformly, independent of the observer or their circumstances.

Albert Einstein's theory of relativity overturned this intuitive understanding. In his special theory of relativity, Einstein proposed that time is not absolute but relative to the observer's frame of reference. Two observers moving at different speeds would measure time differently, a phenomenon known as time dilation. This idea, though counterintuitive, was confirmed by experiments such as those involving atomic clocks on fast-moving planes.

Einstein's general theory of relativity added another layer of complexity, showing that time is influenced by gravity. Clocks run slower in stronger gravitational fields, meaning that time itself is warped by the presence of massive objects. This discovery has profound implications for how we understand the universe, from the behavior of black holes to the expansion of spacetime itself.

Einstein's work forces us to confront a profound truth: time, one of the most familiar aspects of our experience, is not as it seems. Instead, it is a flexible dimension, intertwined with space and shaped by the conditions of the universe.

Challenging the Boundaries of Perception: Heisenberg's Uncertainty Principle

As science delved deeper into the quantum realm, it uncovered truths that defied not only perception but also intuition. Werner Heisenberg's uncertainty principle, one of the cornerstones of quantum mechanics, revealed a fundamental limit to what we can know about the universe. According to this principle, it is impossible to simultaneously know both the position and momentum of a particle with absolute precision. The more accurately we measure one property, the less accurately we can measure the other.

This limitation is not due to technological shortcomings but is inherent in the nature of reality. At the quantum level, particles do not have definite positions or trajectories until they are observed. This challenges classical notions of determinism, where the state of a system at one moment determines its future state.

Implication for Truth:
The uncertainty principle highlights a profound philosophical shift. Truth in the quantum world is not about absolute certainties but probabilities. This realization forces us to rethink what it means to "know" something and raises questions about the role of the observer in shaping reality.

The Blurred Line Between Truth and Interpretation: Schrödinger's Cat

One of the most famous thought experiments in quantum mechanics, Schrödinger's cat, illustrates the tension between truth and perception. In this scenario, a cat is placed in a sealed box with a mechanism that has a 50% chance of releasing a poison, triggered by the decay of a radioactive atom. Until the box is opened and the cat is observed, it is said to exist in a superposition of states—both alive and dead.

This paradox highlights the strange nature of quantum truth. Is the cat truly both alive and dead, or is the superposition merely a limitation of our ability to describe reality? The experiment underscores the role of measurement in defining truth and the challenge of reconciling quantum mechanics with everyday experience.

Conclusion: Truth Beyond Perception

These examples—from the heliocentric revolution to the mysteries of quantum mechanics—demonstrate that truth often lies beyond the reach of our senses and intuition. It requires tools, theories, and the courage to question what seems obvious. Yet, as science pushes the boundaries of perception, it also forces us to confront deeper philosophical questions about the nature of reality and the limits of human understanding.

As we move into the next chapter, we will explore the concept of epistemology—the study of knowledge itself. How do we define knowledge, and how do we navigate its limits in our quest for truth?

The Courage to Change: Scientists Who Reversed Their Theories

Science is often celebrated for its discoveries, but it is equally defined by its capacity for self-correction. Unlike dogmatic systems that cling to certainty, science thrives on its ability to revise and refine its understanding of the world. Some of the greatest minds in history, from Einstein to Darwin, demonstrated intellectual courage by changing their views when confronted with new evidence. This openness to doubt is not a weakness; it is a testament to the strength of scientific inquiry.

Einstein and the Expanding Universe: From Static to Dynamic

Albert Einstein, one of the most celebrated physicists in history, provides one of the most striking examples of a scientist changing his mind. In 1917, Einstein applied his general theory of relativity to the cosmos. His equations suggested that the universe should either be expanding or contracting, but this result conflicted with the prevailing belief that the universe was static and eternal. To reconcile his equations with this widely accepted view, Einstein introduced the cosmological constant—a term that counteracted gravity and kept the universe in equilibrium.

For over a decade, this adjustment seemed consistent with the scientific consensus. However, in 1929, Edwin Hubble's groundbreaking observations shattered the notion of a static universe. Using the Mount Wilson Observatory, Hubble demonstrated that galaxies were moving away from one another, providing the first empirical evidence of an expanding universe. This discovery fundamentally altered our understanding of cosmology, revealing a dynamic and evolving cosmos.

When Einstein learned of Hubble's findings, he reportedly called the introduction of the cosmological constant "the biggest blunder" of his career. Yet this "blunder" was not the end of the story. Decades later, with the discovery of dark energy in the late 20th century, the cosmological constant was reinterpreted as a key component of modern physics, explaining the accelerated expansion of the universe.

Implication for Truth:
Einstein's willingness to revise his views in light of new evidence exemplifies the iterative nature of science. Truth is not static; it evolves with our understanding. This story also highlights the humility required to embrace error, even for the greatest minds.

Charles Darwin: Refining the Mechanisms of Evolution

Another example of a scientist who changed his mind is Charles Darwin, the father of evolutionary theory. Darwin's early work in *On the Origin of Species* emphasized gradualism—the idea that evolutionary changes occur slowly and steadily over long periods. However, as new fossil evidence emerged, Darwin began to reconsider the pace of evolution.

The fossil record revealed periods of relative stability punctuated by sudden bursts of change, a pattern later formalized as "punctuated equilibrium" by Stephen Jay Gould and Niles Eldredge. While Darwin never fully abandoned gradualism, his correspondence reveals an openness to alternative interpretations of the evidence. He acknowledged the gaps in the fossil record and the possibility of more rapid evolutionary processes under certain conditions.

Implication for Truth:
Darwin's flexibility demonstrates that even foundational theories are subject to refinement. His willingness to adapt his views underscores the dynamic nature of scientific truth, which evolves in response to new evidence and perspectives.

Alfred Wegener: From Outcast to Visionary

Alfred Wegener, the German scientist who proposed the theory of continental drift, faced widespread skepticism during his lifetime. His idea—that continents were once connected in a single landmass and drifted apart over time—was dismissed by the scientific community because Wegener could not provide a mechanism to explain how massive landmasses could move.

For decades, geologists clung to the belief that continents were immovable. However, in the mid-20th century, the discovery of plate tectonics provided the missing mechanism, transforming Wegener's once-controversial idea into a cornerstone of modern geology.

Changing Minds in the Face of Evidence:
While Wegener did not live to see his vindication, his work demonstrates the power of evidence to overturn entrenched beliefs. Geologists who had initially dismissed continental drift were forced to change their minds, not out of preference, but because the weight of evidence demanded it.

The Scientific Mindset: A Commitment to Doubt

The stories of Einstein, Darwin, and Wegener highlight a defining feature of science: its commitment to doubt and self-correction. Unlike systems based on dogma or authority, science allows its practitioners to revise their views in light of new evidence.

Feynman's Principle:
Richard Feynman, the Nobel Prize-winning physicist, captured this spirit perfectly: "The first principle is that you must not fool yourself—and you are the easiest person to fool." Feynman's insistence on intellectual honesty reflects the ethos of science. Changing one's mind is not an admission of weakness but a demonstration of strength and integrity.

Modern Implications:
Today, the ability to change one's mind is more important than ever. In fields like climate science, AI ethics, and public health, scientists must remain vigilant, ready to adapt their views as new data emerges. This openness is not a threat to truth; it is the very mechanism by which truth advances.

Conclusion: The Evolution of Truth

The willingness to change one's mind is a hallmark of the scientific process. From Einstein's revision of the cosmological constant to Darwin's reevaluation of evolution, history shows that truth is not a fixed destination but a journey. This capacity for self-correction is what makes science one of humanity's most powerful tools for understanding the universe.

As we continue to explore the philosophy of science, let us remember that doubt is not an obstacle to truth—it is its engine. The next chapter will delve into the limits of human knowledge, exploring how uncertainty shapes the pursuit of truth in science and philosophy.

Science and Dogma: The Struggle Against Certainty

Science is often seen as the antithesis of dogma. Where dogma clings to certainty, science thrives on doubt and inquiry. Yet the history of science reveals that even within its own ranks, dogmatic thinking can emerge, stifling progress and suppressing dissent. The battle between science and dogma is not just a historical narrative but an ongoing struggle, reminding us that the pursuit of truth requires constant vigilance against intellectual complacency.

Galileo and the Inquisition: A Clash of Worldviews

One of the most famous confrontations between science and dogma occurred during the life of Galileo Galilei. In the early 17th century, Galileo used his telescope to make observations that contradicted the geocentric model of the universe, which was deeply intertwined with the religious teachings of the Catholic Church. Galileo's evidence—such as the phases of Venus and the moons of Jupiter—provided strong support for the heliocentric model proposed by Copernicus.

Yet these discoveries were met not with open curiosity but with resistance. The geocentric model was more than a scientific theory; it was a cornerstone of theological and philosophical belief. To challenge it was to challenge the authority of the Church. In 1616, the Church declared the heliocentric model heretical and placed Copernicus's works on the Index of Forbidden Books. Galileo was warned to abandon his support for heliocentrism.

Years later, Galileo published *Dialogue Concerning the Two Chief World Systems*, which presented the heliocentric model as superior to the geocentric one. This work led to his trial by the Inquisition in 1633. Galileo was found guilty of heresy and forced to recant his views under threat of severe punishment. Though he is often quoted as muttering, *"E pur si muove"* ("And yet it moves"), Galileo spent the remainder of his life under house arrest.

Implications for Truth:
The case of Galileo demonstrates how dogma—whether religious, political, or scientific— can act as a barrier to truth. The Church's refusal to engage with new evidence delayed the acceptance of heliocentrism and stifled scientific progress. Galileo's story serves as a cautionary tale about the dangers of conflating authority with truth.

The Lysenko Affair: Dogma in Science

Dogmatic thinking is not limited to religion or philosophy; it can infiltrate science itself. A striking example is the Lysenko affair in the Soviet Union, where ideological dogma suppressed scientific inquiry.

Trofim Lysenko, an agronomist, rejected the principles of Mendelian genetics in favor of a theory that aligned with Marxist ideology. Lysenko argued that environmental conditions, rather than inherited genes, determined the traits of organisms. His ideas were embraced by

Joseph Stalin's regime, as they appeared to support the Marxist belief that human nature and society could be entirely reshaped by external conditions.

Under Lysenko's influence, Soviet agricultural policies abandoned scientific genetics, leading to catastrophic crop failures and widespread famine. Scientists who opposed Lysenko's views were silenced, imprisoned, or executed. One of the most prominent victims was geneticist Nikolai Vavilov, who died in prison for defending Mendelian genetics.

Implications for Truth:
The Lysenko affair reveals how dogma—when paired with political power—can corrupt science and lead to devastating consequences. It underscores the importance of preserving scientific inquiry from ideological interference and fostering an environment where dissent and debate are encouraged.

Dogma in Modern Science: The Danger of Consensus Without Question

While modern science has built mechanisms to guard against dogma, such as peer review and falsifiability, it is not immune to intellectual rigidity.

Example: The Rejection of Plate Tectonics
For much of the early 20th century, the idea of continental drift, proposed by Alfred Wegener, was dismissed by the scientific community. Despite evidence such as the jigsaw-like fit of continents and the distribution of fossils, geologists clung to the belief that continents were fixed in place. This resistance was partly due to the lack of a mechanism to explain how continents could move, but it also reflected a broader reluctance to challenge the prevailing consensus.

It was only in the 1960s, with the discovery of seafloor spreading and the theory of plate tectonics, that Wegener's ideas were vindicated. The shift required scientists to abandon entrenched assumptions and embrace a paradigm that fundamentally redefined geology.

Modern Implication:
Even today, the danger of scientific dogma persists. Fields like climate science, nutrition, and medical research occasionally see the emergence of rigid orthodoxies that resist new ideas. While consensus is important, it must not become a substitute for critical thinking and ongoing investigation.

Philosophical Lessons: The Role of Skepticism in Combating Dogma

Philosophy offers valuable tools for resisting dogma, both within and outside of science. Philosophers like Karl Popper emphasized the importance of falsifiability—the idea that scientific theories must be testable and capable of being proven wrong. Popper argued that dogmatic systems, which shield themselves from criticism, are inherently unscientific.

Example: Newton to Einstein
Isaac Newton's laws of motion and universal gravitation were long considered the pinnacle

of scientific truth. Yet in the early 20th century, Einstein's theory of relativity revealed the limitations of Newtonian physics, showing that its laws were approximations valid only under certain conditions. Einstein's success lay not in discarding Newton's work but in refining it, building a more comprehensive framework.

This transition exemplifies the power of skepticism in advancing knowledge. By questioning even the most established ideas, science continually evolves toward deeper truths.

The Balance Between Certainty and Doubt

While dogma seeks to close the door on inquiry, science keeps it open. Yet, the balance between certainty and doubt is delicate. Too much doubt can paralyze decision-making, while too much certainty can stifle innovation. The strength of science lies in its ability to navigate this tension, embracing uncertainty while striving for reliable knowledge.

Modern Example:
The COVID-19 pandemic highlighted this balance. Early in the pandemic, scientists debated the effectiveness of masks, the origins of the virus, and the best strategies for containment. While the evolving guidance frustrated the public, it reflected the scientific process in action—updating conclusions as new evidence emerged. The eventual development of vaccines showcased science's ability to achieve breakthroughs while grappling with uncertainty.

Conclusion: Guarding Science Against Dogma

The tension between science and dogma is an enduring theme in the history of knowledge. From Galileo's trial to the Lysenko affair, these stories remind us that truth is not guaranteed by authority or consensus but must be continually tested and questioned.

To safeguard science, we must foster a culture of skepticism, encourage dissenting voices, and resist the temptation to equate consensus with certainty. Science's greatest strength lies in its willingness to evolve—and in its refusal to let dogma have the final word.

As we continue our exploration, the next chapter will delve into the limits of human knowledge and the role of epistemology in navigating uncertainty.

Religious Dogma and the Absolute Truth

Throughout human history, religion has been a central framework for understanding existence. From the ancient polytheistic traditions of Egypt and Mesopotamia to the monotheistic faiths of Judaism, Christianity, and Islam, religious systems have sought to answer profound questions about the origin of the universe, the purpose of life, and the nature of morality. These answers, however, are often presented as absolute truths—unchanging, eternal, and beyond question.

Religious dogma is defined by its claim to possess ultimate knowledge, often rooted in sacred texts or divine revelation. This claim is not necessarily a flaw; for many, the certainty offered by religion provides solace and a moral compass. Yet, the insistence on absoluteness can create an intellectual rigidity that resists inquiry, adaptation, or revision. For instance, the Ptolemaic, geocentric model of the cosmos was deeply intertwined with the theological belief that Earth—and by extension, humanity—was at the center of God's creation. Questioning this model was not just a challenge to scientific understanding but an affront to the religious worldview of the time.

The danger of this rigidity becomes apparent when new evidence or ideas emerge that contradict established beliefs. Religious authorities often respond to these challenges not by engaging with them but by dismissing or suppressing them. This pattern can be seen in the trial of Galileo, where the heliocentric model was condemned not on scientific grounds but because it was perceived as heretical, threatening the Church's authority and the coherence of its theological system.

Such resistance to change is rooted in the nature of dogma itself. If a religious doctrine is claimed to be absolute and divinely inspired, admitting error or uncertainty undermines its authority. This has historically led to conflicts between religious institutions and scientific progress, with the former often seeking to enforce conformity and suppress dissent to maintain its claim to ultimate truth.

Science: The Pursuit of Provisional Truths

Science, in contrast, operates on an entirely different epistemological foundation. While religion often claims to start with answers, science begins with questions. Where religious dogma is fixed, science is dynamic, constantly revising its understanding in response to new evidence. This adaptability is one of science's greatest strengths, allowing it to evolve and refine its models of reality over time.

Consider the progression of our understanding of the universe. In the 17th century, Isaac Newton's laws of motion provided a framework that explained planetary motion and the behavior of objects on Earth with extraordinary precision. For centuries, these laws were considered the pinnacle of scientific achievement, a seemingly complete description of the physical world. Yet in the early 20th century, Albert Einstein's theory of relativity revealed that Newtonian mechanics were only an approximation, valid in specific contexts but inadequate at cosmic scales or near the speed of light.

This progression underscores a key difference between science and dogma: science does not claim to have absolute answers. Instead, it seeks the best explanations available given the current evidence, always remaining open to revision. As Carl Sagan famously said, "Science is a way of thinking much more than it is a body of knowledge." Its focus is not on defending existing truths but on uncovering deeper ones, even if that means overturning long-held beliefs.

Epistemological Humility: The Role of Doubt

Another critical distinction between religion and science lies in their relationship with doubt. Religious dogma often views doubt as a threat, a force that undermines faith and corrodes the foundation of belief. In many traditions, questioning divine truths is seen as an act of rebellion or even sin. This perspective has led to the persecution of heretics, the suppression of alternative interpretations, and the marginalization of dissenting voices throughout history.

Science, by contrast, embraces doubt as an essential component of discovery. The scientific method is built on skepticism: hypotheses must be tested, evidence must be scrutinized, and conclusions must be falsifiable. This process ensures that scientific knowledge is not based on authority or tradition but on rigorous inquiry and empirical validation.

Example: The Evolution of Germ Theory
In the 19th century, the idea that diseases were caused by tiny, invisible organisms—bacteria and viruses—was met with skepticism by the medical establishment. The prevailing "miasma theory" attributed disease to bad air or noxious vapors. It took decades of experimentation and evidence, from Louis Pasteur's work on fermentation to Robert Koch's identification of specific pathogens, to establish germ theory as the cornerstone of modern medicine.

This shift required scientists to question deeply ingrained assumptions, often facing resistance from colleagues who were invested in the old paradigm. Yet it also exemplifies how science's embrace of doubt allows it to overcome dogma and arrive at more accurate understandings of the world.

The Social Impact of Dogma and Science

The differing attitudes of religion and science toward truth and doubt have profound social implications. Religious dogma, with its claim to absolute knowledge, often seeks to impose uniformity. This can foster a sense of shared identity and purpose but can also lead to authoritarianism and intolerance. When dogma is wielded as a tool of power, it becomes a mechanism for suppressing dissent and enforcing conformity, as seen in the Spanish Inquisition or the Salem witch trials.

Science, on the other hand, thrives on diversity of thought. Its dynamic nature encourages debate, experimentation, and the exploration of alternative hypotheses. This openness is not just a methodological principle but a cultural one, promoting intellectual freedom and the democratization of knowledge. For example, the open exchange of ideas during the Enlightenment laid the groundwork for revolutions in science, politics, and human rights, challenging both religious and secular dogmas that had constrained progress.

Yet science is not without its dangers. When scientific knowledge is treated as dogma—when it is used to silence dissent rather than to foster dialogue—it can replicate the very rigidity it seeks to avoid. The eugenics movement of the late 19th and early 20th centuries is a cautionary tale, where pseudoscientific ideas about genetics were used to justify racism, forced sterilization, and genocide. This dark chapter reminds us that the pursuit of truth must always be tempered by ethical reflection and humility.

Reconciling Science and Religion

Despite their differences, science and religion are not necessarily irreconcilable. Many religious traditions value inquiry and curiosity, seeing science as a way to understand the divine order of the universe. Figures like Isaac Newton, Gregor Mendel, and Georges Lemaître—the Catholic priest who proposed the Big Bang theory—demonstrate that faith and science can coexist.

However, such reconciliation requires both sides to adopt a posture of humility. Religion must acknowledge the provisional nature of human understanding and resist the temptation to conflate faith with empirical truth. Science, meanwhile, must remain aware of its limits, recognizing that it cannot answer every question, particularly those related to meaning, purpose, or morality.

Conclusion: The Dynamic Nature of Truth

The contrast between religion's dogmatic certainty and science's dynamic inquiry reveals two fundamentally different approaches to truth. While religion often seeks to preserve eternal answers, science is a process of perpetual questioning, refinement, and discovery. Both have their strengths and weaknesses, but the scientific method's embrace of doubt offers a model for navigating the complexities of a changing world.

As we move forward in this exploration, let us remember that truth is not static. Whether revealed through faith or discovered through science, it is a journey rather than a destination, requiring both courage and humility to pursue.

Chapter 5: Scientia et Epistemologia

(Science and Epistemology)

Introduction: What Does It Mean to Know?

What does it mean to know something? This seemingly simple question has perplexed philosophers for centuries and remains central to the philosophy of science. Epistemology—the study of knowledge—asks foundational questions: How do we acquire knowledge? What distinguishes belief from truth? And how can we be certain of anything?

In science, these questions are not merely academic; they shape the very methods by which we investigate the world. From hypotheses and experiments to theories and models, the scientific process relies on assumptions about what can be known and how. Yet, as we delve deeper into the mysteries of the universe, we encounter profound limitations. Can we ever achieve absolute knowledge, or is uncertainty an inherent part of the scientific endeavor?

Plato's Knowledge as Justified True Belief

One of the earliest attempts to define knowledge came from Plato, who described it as *justified true belief*. According to this view, for someone to "know" something, three conditions must be met:

1. The belief must be true.
2. The individual must believe it.
3. The belief must be justified with reasons or evidence.

While this definition seems straightforward, it is not without complications. For example, how do we determine whether a belief is truly justified? And what happens when new evidence overturns long-held beliefs? These challenges reveal the complexity of knowledge, particularly in the context of science, where provisional truths are constantly subject to revision.

Example in Science:
The caloric theory of heat, which posited that heat was a fluid-like substance, was widely accepted in the 18th century. It was consistent with experimental data and justified by prevailing scientific methods. Yet, the development of thermodynamics in the 19th century overturned this theory, showing that heat is a form of energy transfer. This shift underscores the provisional nature of scientific knowledge: what is "justified" today may be disproven tomorrow.

The Problem of Induction: Hume's Challenge

Scottish philosopher David Hume raised a fundamental question about the limits of human knowledge: How can we justify inductive reasoning? Induction—the process of deriving general principles from specific observations—is central to science. For example, we observe that the Sun rises every day and conclude that it will rise again tomorrow.

But Hume argued that this reasoning is circular. We assume that the future will resemble the past because it always has, yet this assumption itself relies on past experience. There is no logical guarantee that the laws of nature will remain constant; we accept them on faith, not certainty.

Scientific Implication:
Hume's problem of induction reveals the inherent uncertainty in scientific knowledge. No matter how many times an experiment is repeated, we can never achieve absolute certainty. This does not diminish the value of science but highlights its provisional nature.

Modern Example:
The discovery of quantum mechanics in the early 20th century challenged long-held assumptions about determinism in physics. Classical mechanics, which had been considered universal, was revealed to be a special case, valid only at macroscopic scales. At the quantum level, particles behave probabilistically, defying the expectations of inductive reasoning based on classical physics.

Karl Popper: Falsifiability as the Criterion of Science

In response to the challenges of induction, philosopher Karl Popper proposed falsifiability as a solution. According to Popper, a scientific theory is not defined by its ability to be proven true but by its capacity to be proven false. For a theory to be scientific, it must make predictions that can be tested and potentially disproven.

Example:
Einstein's theory of general relativity made specific, testable predictions, such as the bending of starlight by gravity. During a 1919 solar eclipse, astronomers confirmed this effect, providing strong evidence for Einstein's theory. Yet the theory remains falsifiable; it could, in principle, be overturned by future observations.

Implications for Epistemology:
Popper's emphasis on falsifiability highlights the provisional nature of scientific knowledge. No theory is ever "proven" in an absolute sense; it survives only as long as it withstands attempts to disprove it. This perspective distinguishes science from dogmatic systems, which resist falsification by shielding their claims from scrutiny.

Thomas Kuhn: Paradigm Shifts and the Structure of Scientific Revolutions

While Popper focused on the falsifiability of individual theories, Thomas Kuhn examined the broader dynamics of scientific progress. In his seminal work, *The Structure of Scientific Revolutions*, Kuhn argued that science does not advance linearly but through periodic "paradigm shifts."

A paradigm is a framework of theories, methods, and assumptions that guide scientific inquiry. For example, Newtonian physics dominated science for centuries, providing a coherent framework for understanding motion and gravity. Yet anomalies—such as the orbit of Mercury, which Newtonian mechanics could not fully explain—accumulated over time, eventually leading to the paradigm shift of Einsteinian relativity.

Implication for Knowledge:
Kuhn's insights reveal that scientific truth is not fixed but context-dependent. Each paradigm offers a different lens through which to view the world, and what is considered "true" within one paradigm may be reinterpreted or discarded in another.

Example: The Germ Theory of Disease
Before the germ theory, diseases were attributed to imbalances in the body's humors or

environmental factors like miasma. The advent of germ theory in the 19th century represented a paradigm shift, fundamentally changing our understanding of medicine and public health. This transformation illustrates how new paradigms reshape not only scientific knowledge but also societal practices.

The Role of Uncertainty in Science

Uncertainty is often viewed as a weakness, yet it is one of science's greatest strengths. Unlike dogmatic systems, which claim absolute certainty, science thrives on questioning its assumptions and embracing ambiguity.

Heisenberg's Uncertainty Principle:
In quantum mechanics, Werner Heisenberg's uncertainty principle reveals a fundamental limit to what we can know about the universe. The more precisely we measure a particle's position, the less precisely we can know its momentum, and vice versa. This limitation is not a result of technological shortcomings but an intrinsic property of nature.

Philosophical Implication:
Heisenberg's work challenges traditional notions of epistemology, suggesting that uncertainty is not merely a lack of knowledge but a defining feature of reality itself. This forces us to reconsider what it means to "know" something in the context of science.

Modern Challenges: The Limits of Human Knowledge

As science pushes the boundaries of understanding, it increasingly confronts questions that may lie beyond the reach of human cognition. Concepts like dark matter, dark energy, and the multiverse challenge not only our scientific tools but also our philosophical frameworks.

Example:
Dark matter and dark energy together account for approximately 95% of the universe's total mass-energy content, yet they remain poorly understood. Despite decades of research, scientists have yet to directly observe or explain these phenomena. This mystery underscores the vastness of what we do not know and may never fully comprehend.

Epistemological Humility:
Such challenges remind us that science is not a path to omniscience but a journey of exploration. Its strength lies not in claiming to have all the answers but in continually seeking better questions.

The Fragility of Certainty: Lessons from the History of Science

Science, despite its achievements, is a fundamentally human endeavor, shaped by the same frailties that govern all intellectual pursuits. The temptation to treat scientific theories as immutable truths has often led to errors that only later generations could rectify. This tension between certainty and revision underscores a profound truth about epistemology: knowledge, even when supported by rigorous evidence, is always provisional.

One of the most telling examples comes from the 19th century, when scientists believed in the existence of a substance called "ether." According to this theory, ether was a medium that permeated all of space, providing the means for light waves to propagate. The existence of ether seemed logical; after all, sound waves required air, so surely light waves required a similar medium. Experiments appeared to confirm its presence, and for decades, the concept of ether was woven into the fabric of scientific thought.

It was Albert Michelson and Edward Morley's famous experiment in 1887 that began to unravel this belief. Their work demonstrated that the speed of light remained constant, regardless of the Earth's motion through space—a result incompatible with the existence of ether. Yet, the scientific community initially struggled to let go of the ether theory. It wasn't until Einstein's theory of special relativity provided a new framework that the idea of ether was fully abandoned. This episode illustrates a critical point: even well-supported scientific theories can crumble when confronted with new evidence, reminding us of the fragility of certainty in science.

The Role of Paradigms in Shaping Knowledge

Thomas Kuhn's concept of paradigms offers a powerful lens for understanding how scientific knowledge evolves. Paradigms are not merely theories; they are comprehensive worldviews that shape how scientists interpret data, conduct experiments, and even frame questions. Within a paradigm, certain truths become so deeply ingrained that they are rarely questioned, creating a sense of stability and coherence. Yet this stability often masks underlying tensions.

For example, during the early 20th century, Newtonian physics provided a seemingly complete description of the physical universe. Its equations could predict the motion of planets, explain the behavior of objects on Earth, and even guide the development of industrial technologies. However, anomalies began to emerge. The orbit of Mercury, for instance, deviated slightly from the predictions of Newtonian mechanics. These discrepancies were initially dismissed as minor errors or the result of unknown factors, but they signaled a deeper problem.

Einstein's general theory of relativity ultimately replaced Newton's framework, revealing that gravity was not a force but the curvature of spacetime. This paradigm shift did not render Newton's laws useless—they remain effective for most practical applications—but it redefined their scope, showing that they were approximations rather than universal truths. Kuhn's insights highlight a key aspect of epistemology: knowledge is not a steady accumulation of facts but a series of revolutions that challenge and transform our understanding of the world.

The Problem of Observer Bias in Scientific Inquiry

Scientific knowledge is often presented as objective, yet the process of discovery is deeply influenced by the biases of those conducting the research. Observer bias occurs when scientists unconsciously interpret data in ways that align with their expectations or assumptions. This bias does not invalidate the scientific method but reminds us of its limitations and the importance of critical scrutiny.

One striking example is the initial rejection of the idea of continental drift, proposed by Alfred Wegener in 1912. Wegener's theory, which suggested that continents had once been connected in a supercontinent called Pangaea, was supported by evidence such as the fit of continental coastlines and the distribution of similar fossils across distant continents. Yet the scientific community largely dismissed his ideas, partly because he lacked a plausible mechanism for how continents could move.

This resistance was not purely based on evidence; it was also shaped by the biases of geologists who were deeply invested in the prevailing view that continents were fixed. It took decades, and the eventual discovery of plate tectonics in the mid-20th century, for Wegener's theory to gain acceptance. This case underscores a fundamental challenge in epistemology: the pursuit of knowledge is always filtered through human perception, which can both illuminate and obscure the truth.

Epistemological Humility: The Importance of Accepting Limits

One of the most profound lessons of epistemology is the recognition that there are limits to what we can know. These limits are not merely practical—such as the difficulty of conducting experiments at extreme scales—but also philosophical, rooted in the very nature of knowledge. Werner Heisenberg's uncertainty principle provides a striking example. In quantum mechanics, it is impossible to simultaneously measure a particle's position and momentum with absolute precision. The more accurately one is measured, the less accurately the other can be known.

This principle is not a result of technological shortcomings but an intrinsic property of nature. At the quantum level, particles do not have fixed positions or velocities until they are observed. This challenges classical notions of determinism, where the future state of a system could, in theory, be predicted with perfect accuracy if its current state were fully known. Instead, the quantum world operates probabilistically, forcing us to rethink what it means to "know" something.

Implications for Science:
The uncertainty principle reveals that there are aspects of reality that lie beyond human comprehension. This does not diminish the value of science but highlights the need for epistemological humility. Science is not a quest for absolute knowledge but an ongoing journey of discovery, where each answer leads to new questions.

The Dynamic Nature of Scientific Models

Scientific models are not meant to be perfect representations of reality; they are tools for understanding and predicting phenomena. Yet the temptation to conflate models with truth is a recurring challenge in epistemology. A historical example is the Ptolemaic system, which placed Earth at the center of the universe. This model, with its intricate system of epicycles, was remarkably effective at predicting the motions of celestial bodies. For centuries, it was treated as an accurate representation of reality.

The Copernican revolution revealed the flaws in this model, yet it also demonstrated the power of models to evolve. Copernicus replaced the geocentric framework with a heliocentric one, and later refinements by Kepler and Newton provided even greater predictive accuracy. Today, our understanding of the cosmos is shaped by Einstein's relativity and the expanding universe model, but even these frameworks are subject to revision as new evidence emerges.

Modern Example:
The current standard model of particle physics is one of the most successful scientific theories ever developed, yet it remains incomplete. It cannot fully explain phenomena such as dark matter and dark energy, which together make up 95% of the universe's total mass-energy content. These mysteries remind us that scientific models are always provisional, approximations that guide us closer to the truth but never capture it entirely.

The Limits of Human Cognition: The Lens Through Which We See Reality

As science seeks to unravel the mysteries of the universe, it becomes increasingly clear that our understanding is shaped—and constrained—by the nature of human cognition. The human brain evolved not as a tool for uncovering universal truths but as a mechanism for survival. Its perceptions, intuitions, and reasoning processes are optimized for navigating the immediate environment rather than comprehending the complexities of quantum mechanics or the vastness of the cosmos. This realization presents a profound epistemological challenge: To what extent can our species, with its biological limitations, truly understand the universe?

Consider the concept of higher dimensions. Our everyday experience is rooted in three spatial dimensions and one temporal dimension, yet modern physics suggests the existence of additional dimensions beyond our perception. Theoretical frameworks like string theory propose that these hidden dimensions are fundamental to the fabric of reality, but they remain inaccessible to direct observation. Mathematicians and physicists use abstract models and equations to describe these dimensions, yet such descriptions are far removed from our intuitive understanding. We can imagine a two-dimensional being attempting to comprehend the third dimension—a challenge akin to our own struggle to conceptualize higher dimensions.

Philosophical Implications:
Immanuel Kant addressed a similar issue in his distinction between phenomena and noumena. According to Kant, we can only know the world as it appears to us (phenomena), filtered through the structures of human perception and cognition. The "thing-in-itself" (noumenon), which exists independently of human experience, remains forever beyond our grasp. This perspective resonates with modern science's acknowledgment of its limitations. While tools like particle accelerators and telescopes extend our senses, they do not transcend the cognitive frameworks through which we interpret data.

Scientific Example:
The quantum world provides a striking example of how cognition struggles to align with reality. Phenomena like wave-particle duality defy the categories of classical physics and human intuition. An electron can behave like a particle or a wave depending on how it is observed—a fact that seems absurd from a macroscopic perspective. This duality is not merely a reflection of limited technology but an intrinsic aspect of quantum systems. The challenge lies not in the quantum world itself but in our inability to perceive or describe it without imposing macroscopic concepts that are fundamentally inadequate.

Cognitive Bias in Science:
The limits of cognition are further compounded by biases inherent in human thought. Confirmation bias, for example, leads scientists to favor evidence that supports their hypotheses while dismissing contradictory data. Even the most rigorous scientific methods cannot entirely eliminate these biases, which influence how experiments are designed, how data is interpreted, and which theories gain prominence. This raises an uncomfortable question: How much of our scientific understanding is shaped not by objective reality but by the constraints of the human mind?

The Pursuit of Knowledge in an Imperfect World

Despite these limitations, science remains humanity's most powerful tool for exploring reality. Its strength lies in its ability to transcend individual biases through collective scrutiny. Peer review, replication, and falsifiability ensure that scientific knowledge is not the product of any one person's cognition but a collaborative effort that iteratively refines our understanding.

Example: The Evolution of Climate Science
The study of climate change illustrates this collective process. Early theories about global warming, based on limited data, faced significant uncertainties. Yet, as more evidence accumulated—from ice core samples to satellite measurements—a coherent picture emerged. This process required the integration of multiple disciplines, including meteorology, oceanography, and physics. While individual researchers inevitably brought their biases and assumptions to the table, the collective nature of the scientific enterprise helped mitigate these limitations, allowing the field to advance.

Science's adaptability to uncertainty is its greatest strength. It does not require perfection or omniscience; it thrives on doubt and imperfection, constantly revising itself in light of new evidence. This stands in stark contrast to systems of thought that demand certainty, reminding us that the pursuit of knowledge is a dynamic, iterative process rather than a destination.

The Illusion of Certainty: When Knowledge Becomes Dogma

One of the recurring challenges in the pursuit of knowledge is the illusion of certainty. Science, despite its dynamic nature, is often misunderstood as a repository of fixed truths rather than an evolving process. This misunderstanding arises partly from the success of scientific theories, which can become so well-established that they are treated as dogmas. Yet history shows that even the most robust scientific ideas are provisional, subject to refinement or even rejection in the light of new evidence.

Consider the case of Newtonian mechanics. By the 18th century, Isaac Newton's laws of motion and universal gravitation had become the foundation of physics. They explained planetary orbits, the behavior of falling objects, and even the tides, with an elegance and precision unmatched by previous theories. For nearly two centuries, Newton's framework was seen as a definitive description of the physical universe, a testament to human ingenuity and reason.

But in the early 20th century, this certainty was shattered. Albert Einstein's theories of special and general relativity revealed that Newton's laws were not universal. While they remain valid for everyday phenomena, they break down at cosmic scales or near the speed of light. Time and space, once thought to be absolute, were shown to be relative, interconnected dimensions shaped by gravity. This paradigm shift was not a rejection of Newton's genius but a recognition that his work was an approximation—a stepping stone toward a deeper understanding of reality.

This lesson is crucial: certainty in science is always an illusion. Knowledge grows not by clinging to the past but by embracing the unknown. The strength of science lies not in its answers but in its willingness to ask better questions.

The Role of Fallibility in Scientific Progress

The recognition of fallibility is one of the most important contributions of epistemology to science. Unlike systems of belief that claim infallibility, science thrives on its ability to identify and correct errors. Every failed experiment, every disproven hypothesis, and every overturned theory contributes to the growth of knowledge.

A powerful example of this principle is the development of evolutionary theory. When Charles Darwin published *On the Origin of Species* in 1859, his ideas were revolutionary, offering a natural explanation for the diversity of life. Yet Darwin's original theory, while groundbreaking, was incomplete. He knew little about the mechanisms of inheritance and struggled to explain how traits were passed from one generation to the next without being diluted.

It was only with the rediscovery of Gregor Mendel's work on genetics in the early 20th century that the puzzle began to be solved. The integration of Mendelian genetics with Darwinian evolution led to the modern synthesis, a more comprehensive framework that united biology, paleontology, and molecular science. Yet even this synthesis is not the final word. Advances in epigenetics and evolutionary developmental biology (evo-devo) continue to refine our understanding of evolution, demonstrating that science is an ever-expanding conversation.

Fallibility is not a weakness; it is the engine of progress. The willingness to admit error and revise beliefs is what allows science to grow, moving closer to truths that are always provisional but increasingly reliable.

Knowledge at the Edge: The Challenge of the Unknown

As science advances, it pushes the boundaries of what can be known. At these edges, the challenges of epistemology become most apparent. The search for dark matter and dark energy, for example, has revealed profound gaps in our understanding of the universe. These mysterious phenomena account for the majority of the universe's mass-energy content, yet they remain undetectable by conventional means.

The existence of dark matter is inferred from its gravitational effects on galaxies, while dark energy is hypothesized to explain the accelerating expansion of the universe. Despite decades of research, their nature remains elusive, leading some scientists to question whether we need new physics to explain these observations. This uncertainty challenges our confidence in established models, forcing us to confront the possibility that our current understanding is not merely incomplete but fundamentally flawed.

Philosophical Implications:
The quest to understand dark matter and dark energy illustrates a deeper philosophical issue: Can human cognition ever fully comprehend the universe? Some thinkers, like Immanuel Kant, argued that certain aspects of reality—the "noumenal world"—are forever beyond our grasp, filtered through the limitations of human perception and reasoning. Modern physics, with its abstractions and paradoxes, seems to confirm this view. The quantum world, for instance, operates according to principles that defy intuition, such as superposition and entanglement. These phenomena challenge not only our understanding but also the very frameworks we use to define knowledge.

The Science of Simplicity and Complexity

One of the paradoxes of science is its simultaneous pursuit of simplicity and complexity. On one hand, scientists seek elegant, unified theories that explain the greatest number of phenomena with the fewest assumptions. Einstein's famous equation, $E=mc2$ $E=mc2$, is a paradigm of simplicity, capturing the equivalence of mass and energy in just five symbols.

On the other hand, the natural world is often irreducibly complex. Biological systems, ecological interactions, and human behavior defy simple explanations, requiring multidisciplinary approaches and probabilistic models. The interplay between simplicity and complexity raises profound epistemological questions: Is simplicity a sign of truth, or is it merely a human preference? How do we balance the need for comprehensible models with the messy realities of the universe?

Example:
The standard model of particle physics is one of the most successful theories in science, accurately predicting the behavior of subatomic particles and forces. Yet it is far from simple.

It relies on a complex framework of mathematical equations and 17 fundamental particles, leaving unanswered questions about gravity and the nature of dark matter. This tension between simplicity and complexity illustrates the dual nature of scientific inquiry, which seeks to make sense of a world that is both orderly and chaotic.

Epistemology and Ethics: The Responsibility of Knowing

The pursuit of knowledge is not merely an intellectual endeavor; it is also an ethical one. Every discovery carries implications, raising questions about how knowledge should be used and who benefits from it. Epistemology challenges us to consider the responsibilities that come with understanding.

Example: The Manhattan Project
The development of nuclear weapons during World War II was a triumph of scientific ingenuity but also a moral dilemma. Scientists like J. Robert Oppenheimer grappled with the ethical implications of their work, recognizing that the knowledge they had unleashed could lead to unprecedented destruction. Oppenheimer's famous reflection, quoting the Bhagavad Gita—*"Now I am become Death, the destroyer of worlds"*—captures the dual-edged nature of scientific progress.

Modern Implications:
Today, advances in artificial intelligence, genetic engineering, and climate science pose similar ethical challenges. As we push the boundaries of knowledge, we must also confront the question: What is our responsibility to future generations? Epistemology reminds us that knowing is not enough; understanding must be paired with wisdom.

From Determinism to Probability: The Epistemological Revolution in Physics

For much of its history, science was guided by the principle of determinism—the belief that the future state of a system could be predicted with absolute precision if its current state were fully known. This view was epitomized by the work of Isaac Newton, whose laws of motion provided a deterministic framework for understanding the universe. According to Newtonian mechanics, every action had an equal and opposite reaction, and the motion of every particle could, in principle, be calculated from its initial conditions. This vision of a clockwork universe was so compelling that Pierre-Simon Laplace famously claimed that a hypothetical intelligence, later called "Laplace's Demon," could predict the entire future of the universe if it knew the position and velocity of every particle.

However, as scientists probed deeper into the nature of matter and energy, cracks began to appear in this deterministic framework. The development of statistical mechanics in the 19th century, led by figures like James Clerk Maxwell and Ludwig Boltzmann, introduced a new way of understanding complex systems. Rather than tracking the precise motion of every molecule in a gas, statistical mechanics focused on probabilities, describing the behavior of the system as a whole. For example, the temperature of a gas could be understood as the average kinetic energy of its molecules, even though the motion of individual molecules was random and unpredictable.

This shift from determinism to probability was not merely a practical solution to the difficulty of tracking countless particles; it represented a fundamental change in how scientists thought about knowledge. No longer was it assumed that every detail of a system could be known with certainty. Instead, knowledge became a matter of statistical inference, grounded in probabilities rather than absolutes.

Epistemological Implications:
The rise of probabilistic thinking challenged the very concept of truth in science. In a deterministic framework, truth is binary: a prediction is either correct or incorrect. In a probabilistic framework, truth becomes a matter of degrees. For example, when meteorologists predict a 70% chance of rain, they are not claiming certainty but expressing a probability based on available data. This probabilistic approach reflects a more nuanced understanding of knowledge, one that acknowledges uncertainty and embraces complexity.

Quantum Mechanics and the Collapse of Determinism:
The probabilistic revolution reached its zenith with the advent of quantum mechanics in the early 20th century. At the quantum level, particles do not follow deterministic trajectories. Instead, their behavior is governed by probabilities, described by a mathematical construct known as the wave function. Erwin Schrödinger's famous equation predicts the likelihood of finding a particle in a particular location or state, but it does not specify the exact outcome of any given observation.

This probabilistic nature of quantum mechanics was famously unsettling to Albert Einstein, who declared, "God does not play dice with the universe." Yet experiments, such as those involving quantum entanglement and the double-slit experiment, confirmed the inherent randomness of the quantum world.

Example: Schrödinger's Cat
Schrödinger's thought experiment illustrates the epistemological dilemma posed by quantum mechanics. In this scenario, a cat placed in a sealed box with a quantum-triggered mechanism is simultaneously alive and dead until observed. This superposition of states highlights the tension between our classical intuitions and the probabilistic nature of quantum reality. What does it mean to "know" the state of the cat if its reality depends on observation?

Beyond Physics: The Spread of Probabilistic Thinking
The influence of probabilistic thinking has extended far beyond physics, transforming fields as diverse as biology, economics, and artificial intelligence. In evolutionary biology, for example, the concept of natural selection is inherently probabilistic, describing how traits increase or decrease in frequency within populations over time. Similarly, economic models use probabilities to predict market behavior, acknowledging the inherent uncertainty of human decision-making.

Even in everyday life, probability has become a central framework for understanding risk and decision-making. From medical diagnoses to weather forecasts, probabilistic reasoning helps us navigate uncertainty in a world that resists deterministic explanations.

Cognitive Resistance to Probability:
Despite its central role in modern science, probabilistic thinking often clashes with human intuition. Studies in cognitive psychology reveal that people struggle to reason probabilistically, frequently overestimating the likelihood of rare events or misinterpreting

statistical data. This resistance highlights a broader epistemological challenge: our cognitive biases and intuitions, shaped by evolution, are often ill-suited to the probabilistic nature of reality.

The Science of Consciousness: Penrose and Hawking's Perspectives

Few questions in science are as profound or as elusive as the nature of consciousness. What is it that gives rise to self-awareness, thought, and subjective experience? Is consciousness simply a byproduct of physical processes in the brain, or does it reflect something deeper about the fabric of reality? The quest to understand consciousness sits at the intersection of science, philosophy, and epistemology, challenging the boundaries of what can be known.

Two of the 20th century's greatest scientific minds, Roger Penrose and Stephen Hawking, approached the question of consciousness from markedly different perspectives. Penrose, with his background in mathematics and an affinity for philosophy, argued that consciousness might lie beyond the reach of classical computation and materialist science. Hawking, with his relentless focus on physics and cosmology, viewed consciousness as a phenomenon that could ultimately be explained within the framework of physical laws. Their debates highlight the tension between reductionist and holistic approaches to understanding the universe—and the limits of scientific knowledge.

Roger Penrose: Consciousness and the Limits of Computation

Roger Penrose, a mathematician and physicist renowned for his work on black holes and general relativity, has long argued that consciousness cannot be fully explained by classical physics or computation. His perspective is rooted in Gödel's incompleteness theorems, which demonstrate that within any sufficiently complex mathematical system, there are truths that cannot be proven using the rules of that system.

Penrose extrapolated this idea to consciousness, suggesting that the human mind is capable of understanding truths that lie beyond the capabilities of algorithmic computation. In his view, consciousness involves non-computable processes that classical physics and artificial intelligence (AI) cannot replicate. He explored this idea in his influential book *The Emperor's New Mind*, proposing that quantum mechanics might hold the key to understanding consciousness.

The Quantum Mind Hypothesis:
Penrose speculated that consciousness arises from quantum processes occurring in the microtubules of brain cells. Microtubules are tiny structures within neurons that play a role in cellular organization and communication. According to Penrose, these structures might facilitate quantum coherence, a state in which particles remain interconnected and act as a unified system, even when separated.

This hypothesis, developed further in collaboration with anesthesiologist Stuart Hameroff, is known as the "Orchestrated Objective Reduction" (Orch-OR) theory. Penrose argued that the collapse of quantum states within microtubules could generate the non-computable processes necessary for conscious thought. While controversial and far from proven, the Orch-OR theory challenges the reductionist view that consciousness is merely the sum of neural activity.

Epistemological Implications:
Penrose's ideas raise profound questions about the nature of knowledge and the limits of computation. If consciousness involves processes that lie beyond algorithmic computation, then our understanding of the mind—and perhaps the universe—requires a fundamentally different framework. This view challenges the prevailing materialist paradigm, suggesting that consciousness may be a fundamental aspect of reality, akin to space, time, and energy.

Stephen Hawking: Consciousness as a Phenomenon of Physics

In contrast to Penrose, Stephen Hawking approached consciousness from a more reductionist perspective. As a physicist, Hawking was deeply committed to the idea that the universe is governed by physical laws and that all phenomena, including consciousness, can ultimately be explained within this framework. While Hawking did not focus extensively on consciousness in his work, his broader views on determinism and the nature of the universe provide insight into his stance.

Materialist Framework:
Hawking viewed the brain as a complex system of neurons and synapses, operating according to the principles of physics and chemistry. In this view, consciousness emerges from the interactions of billions of neurons, with no need for non-computable processes or quantum phenomena. This perspective aligns with the dominant paradigm in neuroscience, which seeks to understand consciousness as a property of the brain that can be studied, modeled, and eventually replicated.

The Role of Complexity:
Hawking acknowledged that consciousness is an extraordinarily complex phenomenon, but he saw this complexity as a challenge for science rather than a barrier to understanding. He believed that advances in neuroscience and computational modeling would eventually unravel the mechanisms underlying thought and awareness. In this sense, Hawking's approach reflected a confidence in the power of reductionist science to explain even the most mysterious aspects of reality.

Epistemological Implications:
Hawking's materialist perspective underscores the role of reductionism in scientific inquiry. By breaking down complex phenomena into their constituent parts, science has achieved remarkable successes, from understanding the behavior of particles to mapping the human genome. Yet this approach also has its limitations, as it struggles to address questions of meaning, subjectivity, and the first-person experience of consciousness.

The Tension Between Reductionism and Holism

The contrasting views of Penrose and Hawking reflect a broader tension in science between reductionist and holistic approaches. Reductionism seeks to explain complex systems by analyzing their components, while holism emphasizes the importance of emergent properties that arise from the interactions of those components. Consciousness, with its intricate interplay of neural, quantum, and subjective elements, epitomizes this tension.

Reductionism's Successes and Limits:
Reductionism has been extraordinarily effective in fields like physics and molecular biology, where the behavior of systems can often be predicted from their parts. However, consciousness resists such explanations. The subjective experience of being—a quality philosophers refer to as "qualia"—cannot be easily reduced to the firing of neurons or the interactions of particles.

Holism's Promise:
Holism offers an alternative, suggesting that consciousness arises from the integration of multiple levels of complexity, from quantum processes to neural networks to social and cultural interactions. Penrose's Orch-OR theory is an example of a holistic approach, seeking to bridge quantum physics and neuroscience to explain the mind. Yet holistic theories often lack the predictive precision of reductionist models, making them harder to test and validate.

Ethical and Philosophical Ramifications

The debate between Penrose and Hawking extends beyond science into philosophy and ethics. If Penrose is correct and consciousness involves processes that transcend materialist explanations, then AI and machine learning systems will never achieve true self-awareness. This perspective has significant implications for debates about the ethical treatment of AI and the limits of artificial intelligence.

Conversely, if Hawking's reductionist view is correct, then consciousness is not unique to humans or even biological systems. In this scenario, creating conscious machines becomes a matter of engineering, raising profound ethical questions about the rights and responsibilities associated with artificial life.

Philosophical Implications:
The nature of consciousness also has implications for epistemology. If Penrose is right, then the limits of computation suggest that human intuition and creativity are unique, irreducible aspects of the mind. If Hawking is correct, then the mind is a product of physical processes, subject to the same laws and constraints as any other system in the universe. These divergent views challenge us to reconsider what it means to know, think, and be.

Penrose's Argument from Gödel's Incompleteness Theorems

At the heart of Roger Penrose's argument about the uniqueness of human consciousness is Kurt Gödel's incompleteness theorems. These theorems, formulated in 1931, demonstrate that any sufficiently complex formal system—such as arithmetic—contains true statements that cannot be proven within the system itself. Gödel's work shattered the dream of a fully axiomatized mathematics, showing that even the most rigorous logical systems have inherent limitations.

Penrose argues that Gödel's theorems have profound implications for the nature of human cognition. Specifically, he contends that the human mind is capable of understanding truths that no algorithmic process can derive. This claim challenges the idea that the brain functions like a computer, processing information through deterministic rules.

Mathematical Foundation:
Gödel's first incompleteness theorem can be expressed as follows:

If a formal system F is consistent (free of contradictions) and capable of expressing basic arithmetic, there exists a statement G in F such that:

1. G is true within the system.
2. G cannot be proven using the axioms and rules of F.

Penrose interprets this result to mean that human mathematicians, using intuition and insight, can recognize the truth of statements like G even though they lie beyond the reach of formal computation. If the human mind were purely algorithmic, it would be bound by the same limitations as Gödel's formal systems.

Example in Mathematics:
Penrose frequently uses the example of the halting problem, a famous result in computer science. Alan Turing showed that it is impossible to write an algorithm that can determine, for all possible inputs, whether another algorithm will eventually halt (stop running) or run forever. This result mirrors Gödel's incompleteness, highlighting the inherent limits of computation. Yet mathematicians routinely solve specific instances of the halting problem through insight and creativity, suggesting that human reasoning transcends algorithmic processes.

Penrose's Quantum Mind Hypothesis

To explain how consciousness might transcend computation, Penrose turns to the field of quantum mechanics. Classical physics, he argues, is deterministic and computational in nature, but quantum mechanics introduces a level of indeterminacy that might underpin the non-computable processes of the mind.

Penrose's hypothesis revolves around two key ideas:

1. **Quantum Superposition:** In quantum systems, particles can exist in multiple states simultaneously until observed. This principle is famously illustrated by Schrödinger's cat, which is simultaneously alive and dead until measured.
2. **Objective Reduction (OR):** Penrose proposes that quantum superpositions collapse not merely due to observation but because of an intrinsic instability governed by gravity. He describes this as "objective reduction" and suggests that this collapse is responsible for generating conscious experience.

The Role of Microtubules:
Collaborating with Stuart Hameroff, Penrose hypothesized that quantum processes occur within microtubules—tiny protein structures in brain cells. These structures, Penrose argues,

might serve as the biological substrate for quantum coherence and objective reduction, giving rise to the non-computable processes necessary for consciousness.

Mathematically, Penrose connects the timescale of quantum collapse to the Planck scale, where gravitational and quantum effects converge. He proposes that the duration of a quantum superposition's collapse is related to the mass displacement of the system:

$$\tau \sim \hbar/E$$

where τ is the collapse time, $\hbar$ is the reduced Planck constant, and E is the gravitational self-energy of the superposition. This equation reflects Penrose's attempt to ground consciousness in fundamental physics, bridging the gap between quantum mechanics and neuroscience.

Critiques of Penrose's View

Penrose's ideas have faced significant criticism, both from neuroscientists and physicists. Critics argue that his interpretation of Gödel's theorems does not necessarily imply that the mind is non-computable. For example:

1. **The Nature of Gödel's Truths:** Philosophers like John Searle and Daniel Dennett contend that Gödel's unprovable truths might not reflect a fundamental difference between human and machine cognition but rather the unique flexibility of biological systems.
2. **Quantum Decoherence:** Many physicists argue that quantum coherence is unlikely to occur in the warm, wet environment of the brain. Quantum effects typically require isolation from the environment, as seen in laboratory experiments involving supercooled particles.

Despite these challenges, Penrose's hypothesis remains influential for its ambition and its willingness to bridge disparate fields. His work invites deeper inquiry into the nature of consciousness and the possibility that it arises from principles not yet fully understood.

Epistemological Implications of Penrose's Ideas

Penrose's perspective forces us to reconsider the epistemological foundations of science and mathematics. If the mind operates on principles that transcend computation, then our understanding of knowledge itself must be redefined. This has profound implications for fields ranging from AI to metaphysics:

1. **Limits of Artificial Intelligence:** If Penrose is correct, AI systems—no matter how advanced—will never achieve true consciousness or creativity. This challenges the view, held by figures like Alan Turing, that human intelligence can be fully replicated by machines.
2. **The Nature of Reality:** Penrose's emphasis on quantum mechanics suggests that consciousness is not merely a product of the brain but an intrinsic feature of the

universe, tied to its fundamental laws. This aligns with philosophical traditions that view mind and matter as interconnected rather than separate.

Conclusion: Knowledge as a Journey, Not a Destination

The philosophy of science reveals that knowledge is not a static collection of facts but a dynamic, evolving process. From the fallibility of human reasoning to the mysteries of the cosmos, the pursuit of knowledge is a journey fraught with uncertainty yet driven by an unrelenting desire to understand.

Epistemology challenges us to embrace the provisional nature of truth, to question our assumptions, and to approach the unknown with humility and curiosity. In the next chapter, we will explore the unseen forces that shape our reality, revealing how science bridges the gap between observation and the theoretical constructs that deepen our understanding of the universe.

Chapter 6: De Causis Invisibilibus

(On Invisible Causes)

Introduction: The Hidden Forces That Shape Reality

Not all truths are visible to the naked eye. The history of science is filled with discoveries that revealed the existence of unseen forces, phenomena that eluded direct observation yet profoundly shaped the world around us. From gravity to electromagnetism, from germs to quantum fields, these invisible causes challenge our intuition and force us to rethink the nature of reality.

This chapter explores the profound implications of invisible forces in science. How do we detect and understand phenomena that we cannot see? What tools and methods allow us to uncover these hidden truths? And what do these forces tell us about the limits of human perception? Through examples spanning history and modern physics, we will examine how science bridges the gap between what is observed and what is real.

Gravity: The Invisible Force That Anchors the Universe

Gravity is one of the most fundamental forces in the universe, yet it is entirely invisible. Its effects—objects falling, planets orbiting stars—are apparent, but its nature remains elusive.

Isaac Newton was the first to formulate a mathematical description of gravity, encapsulated in his law of universal gravitation:

$$F = G \ \frac{m1m2}{r1}$$

Here, FF is the gravitational force between two masses (m1 and m2), r is the distance between their centers, and G is the gravitational constant.

Newton's equations allowed humanity to predict the motion of celestial bodies with extraordinary precision, yet they offered no explanation for how gravity worked. Newton himself acknowledged this limitation, famously writing, *"I frame no hypotheses"* regarding the mechanism of gravity. For centuries, gravity remained a mysterious "action at a distance," an invisible force acting across vast expanses of space.

It was Albert Einstein who revolutionized our understanding of gravity with his general theory of relativity. Einstein proposed that gravity is not a force in the traditional sense but a curvature of spacetime caused by mass and energy. In Einstein's view, massive objects like the Sun create dents in the fabric of spacetime, causing other objects to follow curved paths around them. This paradigm shift transformed our understanding of the universe, yet even Einstein's theory leaves questions unanswered. For example, the nature of gravitational waves—ripples in spacetime first detected in 2015—remains an active area of research.

Electromagnetism: A Hidden World of Fields and Forces

While gravity was the first invisible force to be systematically studied, the discovery of electromagnetism revealed an entirely new realm of unseen phenomena. In the early 19th century, Hans Christian Ørsted and Michael Faraday demonstrated that electricity and magnetism were interconnected. Faraday introduced the concept of fields—regions of space where forces are exerted—to describe how these phenomena could act at a distance.

James Clerk Maxwell unified these ideas in his famous equations, showing that electric and magnetic fields are aspects of a single force: electromagnetism. His equations not only explained existing phenomena but also predicted new ones, such as electromagnetic waves. These waves, which include visible light, radio waves, and X-rays, are all forms of energy traveling through space without requiring a medium.

The discovery of electromagnetic waves fundamentally altered our understanding of the universe. For the first time, scientists realized that light itself is an invisible cause, a vibration of electric and magnetic fields propagating through space. This insight paved the way for technologies like radio, radar, and modern telecommunications, as well as the development of quantum mechanics to explain the behavior of light at microscopic scales.

Germ Theory: Invisible Agents of Disease

Invisible forces are not confined to physics. The discovery of germs—the microscopic organisms that cause disease—marked one of the most important breakthroughs in medical science. Before the germ theory, diseases were often attributed to imbalances in the body or to miasma, "bad air" thought to emanate from decaying matter.

It was Louis Pasteur and Robert Koch in the 19th century who demonstrated that many diseases are caused by specific microorganisms. Using microscopes and experimental methods, they identified bacteria responsible for illnesses like anthrax and tuberculosis. Pasteur's work on fermentation and vaccination further cemented the idea that unseen organisms could have profound effects on human health.

The germ theory not only transformed medicine but also illustrated the power of scientific tools to reveal invisible causes. Through innovations like the microscope, scientists were able to extend human perception, uncovering a hidden world that had shaped human history for millennia.

Quantum Mechanics: The Unseen World of Probability

The quantum revolution of the 20th century unveiled a reality far stranger than anyone had imagined. At the quantum level, particles like electrons and photons behave in ways that defy classical intuition. They can exist in multiple states simultaneously (superposition), "tunnel" through barriers, and become entangled, such that the state of one particle instantly affects another, no matter how far apart they are.

Quantum mechanics describes this behavior using mathematical wave functions, which encode the probabilities of different outcomes. These wave functions are invisible but essential, providing a framework for predicting phenomena that cannot be directly observed. For example, the double-slit experiment reveals that particles act as waves when not observed but as particles when measured, suggesting that observation itself affects reality.

This probabilistic nature of quantum mechanics challenges traditional notions of causality and determinism. While the effects of quantum phenomena are observable—such as in the operation of semiconductors and lasers—the underlying processes remain hidden from direct perception.

Dark Matter and Dark Energy: The Invisible Universe

Perhaps the most profound example of invisible causes in modern science is the discovery of dark matter and dark energy. Together, these phenomena account for approximately 95% of the universe's mass-energy content, yet they remain entirely undetected by conventional means.

Dark Matter:
The existence of dark matter is inferred from its gravitational effects on galaxies. Observations show that galaxies rotate faster than can be explained by visible matter alone, suggesting the presence of an unseen mass. Despite decades of research, dark matter has yet

to be directly observed or identified, leaving its nature one of the greatest mysteries in cosmology.

Dark Energy:
Dark energy, on the other hand, is hypothesized to explain the accelerating expansion of the universe. Observations of distant supernovae in the 1990s revealed that galaxies are moving apart at an increasing rate, contrary to expectations. Dark energy is thought to be a form of energy inherent to spacetime itself, yet its properties remain speculative.

The study of dark matter and dark energy illustrates the limitations of human perception and the power of indirect inference. By observing their effects on the cosmos, scientists glimpse the contours of a reality that lies beyond direct observation.

A World Beyond Sight

These examples—gravity, electromagnetism, germs, quantum mechanics, and dark matter—demonstrate the profound role of invisible causes in shaping our understanding of the universe. They challenge our reliance on sensory perception and highlight the tools and theories that allow science to extend its reach into the unseen.

The next section will delve deeper into the philosophical implications of these discoveries, exploring how the recognition of invisible causes has reshaped our understanding of truth, causality, and the limits of human knowledge.

The Evolution of the Atomic Model: From Orbits to Waves

The history of the atomic model reflects humanity's deepening understanding of the unseen forces and structures that govern reality. Early in the 20th century, the structure of the atom was largely a mystery. Scientists knew atoms consisted of a positively charged nucleus surrounded by negatively charged electrons, but the nature of these electrons' motion was unclear. The model presented in this diagram captures the profound shift from classical to quantum mechanical descriptions of the atom, illustrating how science moves from observable phenomena to more abstract, yet accurate, representations of reality.

On the left side of the diagram, the depiction of electrons moving in fixed orbits around a nucleus reflects the **Bohr model of the atom** (1913). Niels Bohr refined Ernest Rutherford's earlier model by incorporating quantum ideas. Bohr proposed that electrons occupy discrete energy levels ($N=1,2,3,\ldots N=1,2,3,\ldots$), where they move in circular orbits without radiating energy. These orbits correspond to specific energy states, and electrons can jump between levels by absorbing or emitting a photon of light with energy equal to the difference between these levels. This model successfully explained the spectral lines of hydrogen, providing one of the first glimpses of quantum behavior in action.

However, as experimental techniques advanced, limitations in the Bohr model became apparent. The orbits depicted in the left part of the diagram are an oversimplification. Electrons do not follow fixed circular paths but instead exhibit behaviors that can only be described probabilistically. This realization gave rise to the **quantum mechanical model of the atom**, depicted on the right side of the diagram. Here, the electron is no longer a particle confined to a single path but is described as a wave, with its behavior determined by a wavefunction. This shift represents a profound departure from classical intuition, replacing certainty with probability.

Wave Patterns and the Quantum Nature of Electrons

The right-hand side of the diagram illustrates the concept of **wave-particle duality**, a cornerstone of quantum mechanics. Electrons exhibit both particle-like and wave-like properties, depending on how they are observed. The wave patterns in the diagram represent the probability distributions of an electron's position, commonly referred to as orbitals. These patterns are determined by solving Schrödinger's wave equation, which describes the quantum state of a system.

For example, the circular wave patterns labeled $N=4N=4$ show that the electron's position is not fixed but is distributed in regions where the wave amplitude is highest. These regions correspond to higher probabilities of finding the electron. This probabilistic nature is a radical departure from classical physics, where the trajectory of a particle is always precisely defined. Instead, in quantum mechanics, we deal with probabilities and uncertainties.

The dashed and solid lines in the diagram highlight different possible wave patterns (e.g., **wave pattern 1** and **wave pattern 2**). These patterns reflect the fact that electrons can exist in multiple quantum states simultaneously until measured, a phenomenon known as **superposition**. This wave-based representation not only explains the behavior of electrons in atoms but also underpins technologies like lasers, semiconductors, and quantum computing.

Models of atomic structure

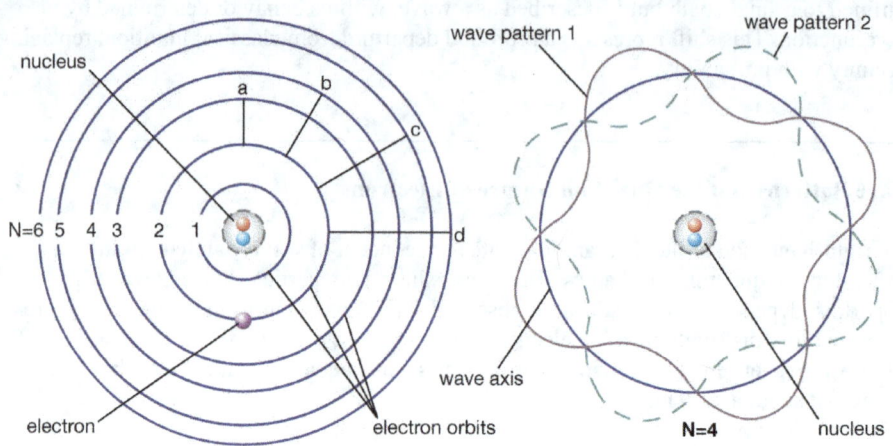

© 2014 Encyclopædia Britannica, Inc.

Commentary on the Diagram: The Transition from Classical to Quantum Thinking

The contrast between the left and right sides of the diagram encapsulates the scientific revolution that occurred in the early 20th century. The Bohr model on the left represents the final stage of classical thinking, where electrons are treated as particles orbiting a nucleus, much like planets orbit a star. While this model was a step forward, it could not explain more complex phenomena, such as the spectra of heavier atoms or the fine structure of spectral lines.

The right-hand side of the diagram represents the modern quantum mechanical model, which supersedes the Bohr model by incorporating the wave-like behavior of electrons. This shift required abandoning classical notions of certainty and determinism. Electrons are no longer particles with precise locations and velocities; instead, they are described by wavefunctions that encode the probabilities of various outcomes. This probabilistic nature reflects the **Heisenberg uncertainty principle**, which states that we cannot simultaneously know both the position and momentum of an electron with perfect accuracy.

The move from orbits to wave patterns also reflects a deeper philosophical shift. The classical Bohr model assumes that reality is directly observable and deterministic, while the quantum mechanical model acknowledges the limits of observation and embraces the probabilistic nature of the universe. This shift has profound implications for epistemology, forcing us to rethink what it means to "know" something when certainty is unattainable.

Epistemological Implications of the Quantum Model

The quantum mechanical model raises profound questions about the nature of reality and our ability to comprehend it. If electrons exist as waves of probability, what does this say about the nature of matter itself? Is reality fundamentally indeterminate, or is our uncertainty a reflection of our limited observational tools? These questions have sparked intense debate among scientists and philosophers alike.

For Roger Penrose, the quantum nature of electrons suggests a deep connection between the physical universe and consciousness. He argues that the collapse of the wavefunction—the process by which an electron's probabilistic wave becomes a definite state—may hold the key to understanding the mind. Penrose's view challenges the reductionist assumption that consciousness can be fully explained by classical physics, suggesting that quantum processes might play a role in the emergence of awareness.

Stephen Hawking, on the other hand, viewed the quantum model as a testament to the power of physical laws. While he acknowledged the probabilistic nature of quantum mechanics, Hawking remained committed to the idea that science could eventually provide a complete understanding of the universe, including consciousness. This tension between reductionism and holism reflects the ongoing challenge of integrating quantum mechanics with other aspects of human experience.

From Atoms to the Cosmos: The Broader Significance of Invisible Causes

The quantum mechanical model, as depicted in this diagram, is not just a representation of the atom; it is a window into the underlying structure of reality. The wave patterns and probability distributions remind us that the universe is far more complex—and far more mysterious—than our senses can perceive. The fact that electrons, the building blocks of matter, defy classical intuition highlights the limitations of human cognition and the power of scientific models to expand our understanding.

The discoveries represented here are not isolated to physics; they echo across disciplines. The probabilistic nature of quantum mechanics has inspired new approaches in fields as diverse as biology, economics, and artificial intelligence. The tools and theories developed to study electrons are now being used to explore the fundamental questions of life, society, and consciousness.

The Invisible Core: Insights from Nuclear Physics

At the heart of every atom lies the nucleus, a dense core of protons and neutrons bound together by one of nature's most powerful invisible forces: the strong nuclear force. Discovered in the early 20th century, the atomic nucleus is far smaller than the atom as a whole, yet it contains almost all of its mass. The discovery of the nucleus and the forces that govern it opened a new chapter in physics, revealing not only the structure of matter but also the energy locked within it—a revelation that would transform both science and society.

The journey to understanding the nucleus began with experiments like those conducted by Ernest Rutherford in 1911. Rutherford's gold foil experiment demonstrated that atoms are mostly empty space, with a tiny, positively charged nucleus at their center. This was a revolutionary finding, as it contradicted earlier models that envisioned atoms as uniform spheres of matter. The nuclear model of the atom not only redefined our understanding of matter but also posed new questions: What holds the nucleus together despite the repulsive forces between positively charged protons? And what governs the interactions between its components?

The answers to these questions came with the discovery of the **strong nuclear force**, a fundamental interaction that binds protons and neutrons together. Unlike gravity or electromagnetism, the strong force operates only at extremely short distances, on the scale of the nucleus itself. It is this force that prevents the positively charged protons from flying apart due to their mutual repulsion. The discovery of the strong force—and its counterpart, the weak nuclear force, which governs processes like radioactive decay—revealed a hidden layer of reality, where forces far stronger than gravity or electromagnetism dominate.

The Philosophical Implications of Nuclear Physics

Nuclear physics challenges many of our intuitions about the nature of reality. At the macroscopic level, the objects we encounter appear solid and stable. Yet nuclear physics reveals that this stability is an illusion, the result of delicate balances between competing forces at the subatomic level. Without the strong nuclear force, matter as we know it could not exist; the universe would consist only of free-floating protons and neutrons.

This realization underscores the theme of **invisible causes** that runs through this chapter. The stability of matter, the energy released in nuclear reactions, and the processes that power stars all depend on forces that are entirely imperceptible to human senses. These forces are inferred through their effects—such as the binding energy of nuclei or the light emitted by the Sun—and described through mathematical models. The fact that we can understand these invisible forces so precisely speaks to the power of science to extend human perception beyond its natural limits.

Philosophically, nuclear physics also raises questions about reductionism and holism. Reductionism seeks to explain complex phenomena by breaking them down into their constituent parts, a method that has been remarkably successful in nuclear physics. By studying individual protons, neutrons, and the interactions between them, physicists have developed detailed models of nuclear structure and behavior. Yet the complexity of nuclear systems also highlights the limitations of reductionism. For example, the collective behavior of particles within a nucleus often cannot be predicted by studying individual particles in isolation. Instead, emergent properties—such as the energy levels of nuclei or the processes that occur in nuclear fusion—arise from the interactions between particles, requiring a more holistic approach to understanding.

The Energy Within: Nuclear Reactions and the Hidden Power of the Atom

Nuclear physics is not just about understanding the structure of matter; it is also about harnessing the energy contained within the nucleus. The discovery that mass and energy are interchangeable, as expressed in Einstein's famous equation $E=mc2E=mc2$, revealed the staggering potential of nuclear energy. Even a tiny amount of mass can be converted into an enormous amount of energy, as seen in processes like nuclear fission and fusion.

In nuclear fission, the nucleus of a heavy atom, such as uranium or plutonium, splits into smaller nuclei, releasing energy in the process. This energy comes from the difference in binding energy between the original nucleus and the resulting fragments. The practical applications of fission, from nuclear power plants to atomic weapons, have had profound social and ethical implications, forcing humanity to grapple with the dual-edged nature of scientific discovery.

Nuclear fusion, on the other hand, is the process that powers stars. In the Sun's core, hydrogen nuclei fuse to form helium, releasing energy that sustains life on Earth. Fusion represents a cleaner and potentially limitless source of energy, but replicating the conditions required for fusion—extreme temperatures and pressures—remains a formidable scientific challenge. These processes illustrate how the invisible forces within the nucleus shape the macroscopic world, from the light we see to the energy we use.

Nuclear Physics and the Philosophy of Science

The development of nuclear physics also highlights key themes in the **philosophy of science**, particularly the relationship between theory and observation. Many aspects of nuclear physics, from the behavior of quarks to the mechanisms of nuclear decay, cannot be directly observed. Instead, they are inferred through experiments and described using abstract mathematical models. This reliance on indirect evidence raises questions about the nature of scientific truth. Are these models accurate representations of reality, or are they merely tools for predicting phenomena?

One of the most significant philosophical debates in nuclear physics concerns the interpretation of quantum mechanics, which underpins our understanding of the nucleus. The behavior of particles within the nucleus is governed by quantum principles, such as superposition and entanglement, which challenge classical notions of causality and determinism. For example, in radioactive decay, a nucleus transitions from one state to another in a probabilistic manner, with no way to predict exactly when the decay will occur. This unpredictability forces us to confront the limits of human knowledge and the possibility that the universe is fundamentally indeterminate.

Another philosophical issue raised by nuclear physics is the ethical responsibility of science. The same principles that explain the forces within the nucleus have been used to create weapons of unparalleled destructive power. The bombings of Hiroshima and Nagasaki in 1945 demonstrated the devastating potential of nuclear physics, raising questions about the role of scientists in shaping the consequences of their discoveries. Can science ever be value-neutral, or does the pursuit of knowledge carry inherent ethical obligations? These questions remain as relevant today as they were in the aftermath of World War II.

Nuclear Forces and the Invisible Threads of the Universe

The study of nuclear physics reveals a universe woven together by invisible threads. The strong nuclear force binds protons and neutrons within the nucleus, while the weak nuclear force governs processes like radioactive decay, enabling the formation of new elements in stars. These forces, though imperceptible, shape the macroscopic world in profound ways.

At the same time, nuclear physics forces us to grapple with the limitations of human perception and understanding. The nucleus, with its tiny size and immense energy, exemplifies the challenge of studying phenomena that lie beyond the reach of our senses. Yet it also demonstrates the power of science to bridge this gap, using tools, theories, and models to uncover the hidden forces that govern reality.

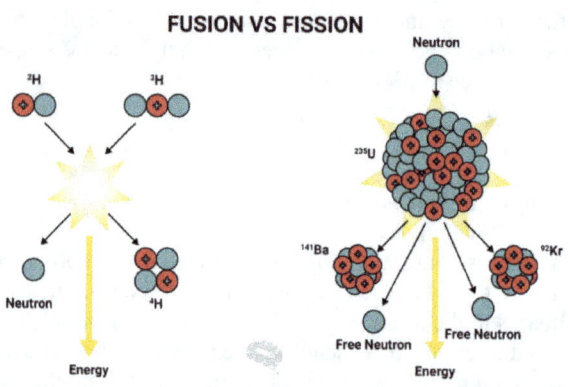

The Two Faces of Nuclear Energy: Fission and Fusion

The diagram illustrates two fundamental processes in nuclear physics—**fission** and **fusion**—which are responsible for some of the most powerful forces in the universe. While these phenomena occur on a scale far smaller than we can observe directly, their effects ripple out to shape the cosmos and human civilization alike. Both processes involve the nucleus, the dense heart of the atom, and both release extraordinary amounts of energy, yet they differ profoundly in their mechanisms, applications, and implications for the future of science and humanity.

Nuclear Fission: Breaking Apart the Atom

On the left side of the diagram, we see nuclear fission, the process by which a heavy atomic nucleus splits into smaller nuclei, accompanied by the release of energy and free neutrons. The sequence begins when a neutron strikes the nucleus of a heavy element, such as uranium-235 or plutonium-239. This collision destabilizes the nucleus, causing it to split into two lighter nuclei, often referred to as "fission fragments." This process releases a significant amount of energy, which comes from the conversion of mass into energy, as described by Einstein's equation $E=mc^2$.

Fission also liberates additional neutrons, which can trigger a **chain reaction**. In a controlled environment, such as a nuclear reactor, this chain reaction can be harnessed to generate electricity. The heat produced by fission is used to boil water, creating steam that drives turbines to produce power. However, if the chain reaction is uncontrolled, as in a nuclear weapon, the result is an explosive release of energy, capable of devastating entire cities.

Applications and Ethical Dilemmas:

The discovery of fission marked a turning point in human history, unlocking a source of energy millions of times more efficient than chemical reactions. Yet it also posed profound ethical questions. The development of atomic bombs during World War II demonstrated both the potential and the peril of harnessing the atom's power. The bombings of Hiroshima and Nagasaki not only ended the war but also ushered in an age of existential risk, where humanity gained the capability to destroy itself.

Philosophically, fission raises questions about the dual-edged nature of scientific discovery. Can knowledge ever be pursued without considering its consequences? Does the potential for harm outweigh the benefits of progress? These questions remain central to debates about nuclear energy and weapons proliferation today.

Nuclear Fusion: The Power of Creation

On the right side of the diagram, we see nuclear fusion, the process by which light atomic nuclei combine to form a heavier nucleus, releasing energy in the process. Fusion occurs when two isotopes of hydrogen, such as deuterium and tritium, collide with enough force to overcome the electrostatic repulsion between their positively charged protons. When these nuclei fuse, they form helium and release a neutron, along with a vast amount of energy.

Fusion is the process that powers the Sun and other stars, where extreme temperatures and pressures in their cores provide the conditions necessary for fusion to occur. On Earth, replicating these conditions has proven to be one of the greatest scientific and engineering challenges. Experimental reactors like tokamaks and stellarators aim to confine superheated plasma using powerful magnetic fields, but achieving sustained fusion has so far remained elusive. Despite these challenges, fusion holds the promise of virtually limitless energy with minimal environmental impact, as it produces no long-lived radioactive waste and relies on abundant fuel sources like hydrogen.

Applications and Philosophical Implications:

Fusion represents the opposite face of nuclear energy compared to fission. While fission is often associated with destruction, fusion symbolizes creation. It is the force that drives the stars and forges the elements, from the hydrogen that fuels stars to the carbon and oxygen that make life on Earth possible. Harnessing fusion would be a monumental achievement, offering humanity a clean and sustainable energy source.

Philosophically, fusion represents the human aspiration to transcend our limitations and emulate the processes of the cosmos. Yet it also forces us to confront our technological and ethical boundaries. If we achieve controlled fusion, will it be used solely for peaceful purposes, or will it become yet another tool for destruction? The hydrogen bomb, which relies on fusion, demonstrates the potential for both creation and devastation inherent in this process.

Debating Fission vs. Fusion: Energy, Risk, and the Future

Fission and fusion represent two sides of the same coin: the manipulation of the atomic nucleus to release energy. However, they differ fundamentally in their risks, rewards, and philosophical significance.

Energy Density and Efficiency:
Both processes release enormous amounts of energy compared to chemical reactions, but fusion is far more efficient. The fusion of a single gram of hydrogen can release as much energy as burning thousands of kilograms of coal. This efficiency makes fusion a tantalizing prospect for solving the world's energy needs. However, fission remains more practical with current technology, as it can be controlled and sustained in reactors.

Safety and Waste:
Fission, while a proven technology, produces radioactive waste that remains hazardous for thousands of years. It also carries the risk of catastrophic accidents, such as those at Chernobyl and Fukushima. Fusion, on the other hand, produces no long-lived waste and is inherently safer, as a failure in a fusion reactor would simply cause the reaction to stop. However, achieving the conditions necessary for fusion is an immense technical challenge, requiring temperatures hotter than the Sun's core and advanced containment systems.

Philosophical Considerations:
Fission exemplifies the power of science to break things apart—to deconstruct and analyze the components of nature. It aligns with a reductionist approach, where understanding the parts of a system leads to control over the whole. Fusion, by contrast, reflects a holistic view, emphasizing the creative potential of science to unite and build. The two processes symbolize the duality of human ingenuity: our capacity to harness the forces of nature for both destruction and creation.

Fission also raises ethical questions about the unintended consequences of scientific progress. The development of nuclear weapons, while grounded in the pursuit of knowledge, has forced humanity to live under the shadow of potential annihilation. Fusion, meanwhile, represents the hope that science can be a force for good, addressing global challenges like climate change and energy poverty.

Invisible Causes in the Nuclear Age

Both fission and fusion illustrate the power of invisible forces to shape our world. The energy released in these processes comes from the binding energy that holds the nucleus together, a force imperceptible to the senses yet fundamental to the structure of matter. Understanding and harnessing these forces requires tools and theories that extend far beyond human perception, from particle accelerators to quantum field equations.

At the same time, these processes challenge us to think deeply about the role of science in society. Should we pursue knowledge for its own sake, or must we always consider its implications? How do we balance the risks of scientific progress against its potential benefits? Fission and fusion are not just physical phenomena; they are windows into the ethical and philosophical dilemmas that define the human condition.

Radioactivity: The Unseen Transformation of Matter

One of the most profound discoveries of the late 19th century was the phenomenon of radioactivity—the spontaneous emission of energy and particles from the nucleus of an unstable atom. This invisible process, first observed by Henri Becquerel and later studied in detail by Marie and Pierre Curie, revealed a previously unknown form of transformation occurring at the atomic level. Radioactivity not only revolutionized our understanding of matter but also provided humanity with powerful tools for medicine, energy, and industry.

Radioactive decay occurs when an unstable atomic nucleus releases energy to become more stable. There are three primary types of radioactive emissions: **alpha particles**, which are positively charged helium nuclei; **beta particles**, which are high-energy electrons or positrons; and **gamma rays**, which are electromagnetic waves of extraordinarily high energy. These emissions are entirely invisible to human senses but can be detected through their effects, such as ionizing radiation or damage to biological tissue.

The discovery of radioactivity shattered the classical view of the atom as an indivisible unit of matter. Instead, it revealed that atoms are dynamic systems, capable of undergoing fundamental changes. This realization had profound implications, leading to the development of nuclear physics and opening new avenues for exploration, such as the age of the Earth (through radiometric dating) and the inner workings of stars.

Philosophical Implications:
Radioactivity exemplifies the theme of **invisible causes** by showing that fundamental processes often occur beyond the realm of human perception. The transformation of one element into another through radioactive decay challenges classical notions of permanence and stability, emphasizing the dynamic nature of matter. Philosophically, this discovery raises questions about the nature of time and change. If atoms themselves are not eternal, what does this say about the permanence of the universe as a whole?

On a practical level, radioactivity also forces us to grapple with ethical dilemmas. The same process that powers life-saving medical treatments, such as cancer radiotherapy, can also cause severe harm, as evidenced by the effects of radiation exposure in nuclear disasters like Chernobyl and Fukushima. These dual aspects of radioactivity—its capacity to heal and to harm—highlight the complex relationship between science and society.

Neutrinos: The Ghost Particles of the Universe

Another example of an invisible cause with far-reaching implications is the neutrino, a subatomic particle so elusive that it has been nicknamed the "ghost particle." Neutrinos were first theorized by Wolfgang Pauli in 1930 to account for the apparent loss of energy in beta decay, a process in which a neutron transforms into a proton while emitting an electron. Pauli proposed that an additional, nearly undetectable particle was carrying away the missing energy, preserving the fundamental principle of conservation of energy.

Neutrinos are electrically neutral and interact with matter so weakly that billions of them pass through your body every second without leaving a trace. Detecting neutrinos requires extraordinary experimental setups, such as massive underground detectors filled with water or heavy liquids, shielded from other forms of radiation. The confirmation of neutrinos in the mid-20th century was a triumph of theoretical physics, providing indirect evidence for an invisible yet essential component of the universe.

Astrophysical Significance:
Neutrinos play a crucial role in astrophysics, particularly in the study of supernovae and the life cycles of stars. When a massive star explodes as a supernova, it releases a burst of neutrinos that carry information about the extreme conditions within the collapsing core. Detecting these neutrinos allows scientists to peer into processes that are otherwise hidden from view, offering insights into the formation of neutron stars and black holes.

The Role of Neutrinos in Cosmology:
In addition to their role in stellar phenomena, neutrinos are key to understanding the evolution of the universe. They were produced in vast quantities during the first moments after the Big Bang and continue to permeate space as part of the cosmic neutrino background. Studying these primordial neutrinos could shed light on the earliest moments of the cosmos, providing clues about the nature of dark matter, the asymmetry between matter and antimatter, and the fundamental laws of physics.

Philosophical Implications:
The discovery and study of neutrinos challenge traditional notions of visibility and detectability in science. These particles exist at the edge of what can be observed, highlighting the limits of human perception and the ingenuity required to explore the unseen. Their weak interactions with matter also raise questions about the nature of existence itself. If most of the universe is composed of entities like neutrinos and dark matter—imperceptible to our senses—how do we define what is "real"?

Furthermore, neutrinos underscore the importance of indirect evidence in science. Just as gravity is inferred from its effects on objects, neutrinos are understood through their interactions, however rare. This reliance on inference rather than direct observation reflects the broader epistemological challenges of modern science, where much of what we "know" about the universe comes from phenomena that cannot be directly seen or touched.

The Philosophy of Invisible Causes: A Bridge Between the Seen and Unseen

Both radioactivity and neutrinos exemplify the power of science to uncover the unseen forces that shape reality. These phenomena remind us that human perception is limited, and that the universe operates on scales and in ways far beyond our immediate experience. Yet, through ingenuity and perseverance, scientists have developed tools and theories that allow us to infer the existence of these invisible causes, bridging the gap between the seen and the unseen.

At the same time, these discoveries challenge our philosophical assumptions about knowledge and reality. If much of what shapes the universe lies beyond direct observation, what does this say about the nature of truth? Are scientific models approximations of reality, or do they capture something fundamental? And how do we reconcile the power of invisible forces with the limitations of human understanding?

In addressing these questions, the study of invisible causes highlights the dynamic and provisional nature of science. Far from providing fixed answers, science is a process of constant questioning, refinement, and discovery. By exploring phenomena like radioactivity and neutrinos, we are reminded that the search for truth is as much about the mysteries we uncover as it is about the questions that remain.

Chapter 7: De Fallaciis Mentis Humanae

(On the Fallacies of the Human Mind)

Introduction: The Mind as a Lens, and Its Distortions

The human mind is our greatest tool for understanding the universe. It has unlocked the secrets of the atom, mapped the stars, and deciphered the codes of life itself. Yet the same mind is also prone to error, bias, and distortion. As much as it seeks truth, it is often led astray by fallacies rooted in perception, cognition, and emotion.

This chapter explores the fallacies of the human mind—those cognitive biases, shortcuts, and errors that can hinder scientific progress and obscure philosophical inquiry. By examining how these fallacies operate, we can better understand the limits of human cognition and the importance of rigorous methods in the pursuit of knowledge. From the overconfidence of brilliant minds to the collective delusions of entire societies, this chapter reveals how the mind can deceive itself—and how science and philosophy can help us overcome these deceptions.

The Fallacy of Intuition: When Instinct Misleads

Human intuition is a powerful tool, honed by evolution to help us navigate a complex world. It allows us to make quick decisions and recognize patterns, often without conscious thought. Yet intuition is not always reliable. In fact, it can lead us astray in situations where reality defies our expectations.

Example: The Flat Earth Fallacy

One of the earliest and most persistent examples of intuitive error is the belief that the Earth is flat. To the unaided senses, the ground appears flat, and the horizon seems to stretch infinitely in all directions. For much of human history, this intuitive perception shaped our understanding of the world. It was only through careful observation and reasoning—such as

Eratosthenes' measurement of the Earth's circumference in ancient Greece—that the true shape of the Earth was revealed.

Even today, flat-Earth conspiracies persist, fueled by the human tendency to trust intuition over evidence. This highlights a broader challenge: our intuitions are often shaped by limited perspectives and fail to account for the complexities of reality. Science, with its emphasis on measurement and falsifiability, provides a corrective lens, allowing us to see beyond the constraints of intuition.

Confirmation Bias: Seeing What We Want to See

Another common fallacy is confirmation bias, the tendency to seek out and favor information that confirms our existing beliefs while ignoring or dismissing evidence that contradicts them. This bias is particularly insidious because it operates below the level of conscious awareness, subtly shaping how we perceive and interpret the world.

Example: Einstein and the Cosmological Constant

Even the greatest minds are not immune to confirmation bias. When Albert Einstein formulated his general theory of relativity, his equations suggested that the universe should be either expanding or contracting. However, this result conflicted with the prevailing belief that the universe was static. To reconcile his theory with this assumption, Einstein introduced the **cosmological constant**, a term that counteracted gravity and maintained a static universe.

Decades later, Edwin Hubble's observations revealed that the universe is indeed expanding, leading Einstein to famously call the cosmological constant his "greatest blunder." This episode illustrates how even a brilliant scientist like Einstein could be influenced by the biases of his time, shaping his interpretation of evidence.

Overconfidence: The Hubris of Certainty

Overconfidence is one of the most dangerous cognitive fallacies, particularly in the context of science and philosophy. It occurs when individuals or societies place undue certainty in their beliefs, assuming they have arrived at ultimate truths when, in fact, they have only scratched the surface.

Example: The Luminiferous Ether

For centuries, physicists believed in the existence of the luminiferous ether, a hypothetical medium through which light waves were thought to propagate. The ether was considered so self-evident that it became an unquestioned part of scientific theory. It was only with the experiments of Michelson and Morley in the late 19th century, and the subsequent development of Einstein's special relativity, that the ether was discarded.

The collapse of the ether theory serves as a cautionary tale about the dangers of overconfidence. It reminds us that even the most widely accepted ideas must remain open to scrutiny, and that scientific progress depends on a willingness to question foundational assumptions.

The Gambler's Fallacy: Misjudging Randomness

The human mind has a natural tendency to seek patterns, even in random events. This tendency, while useful for survival, often leads to errors in reasoning. The **gambler's fallacy** is a classic example: the belief that if an event has occurred frequently in the past, it is less likely to occur in the future, or vice versa. For instance, after a long streak of coin tosses resulting in heads, people often assume that tails is "due," even though each toss is independent.

Implications in Science and Decision-Making:

The gambler's fallacy highlights a broader problem in how humans perceive probability and randomness. In fields like medicine, finance, and climate science, misjudging randomness can lead to faulty predictions and poor decisions. Recognizing this fallacy is essential for developing more accurate models and avoiding costly mistakes.

The Collective Fallacy: Groupthink and the Illusion of Consensus

Human beings are social creatures, and our beliefs are often shaped by the groups to which we belong. While collective thinking can lead to powerful collaborations, it can also give rise to **groupthink**, a phenomenon where the desire for consensus suppresses dissent and critical thinking. Groupthink can result in flawed decisions, as individuals conform to the majority view rather than voicing their own perspectives.

Example: Phlogiston Theory

In the 17th and 18th centuries, the phlogiston theory dominated chemistry. According to this theory, combustion and rusting were caused by the release of a substance called phlogiston. Despite mounting evidence against it, the theory persisted because it was widely accepted by the scientific community. It was only with the work of Antoine Lavoisier, who demonstrated the role of oxygen in combustion, that phlogiston theory was finally overturned.

This example illustrates how groupthink can delay scientific progress by suppressing alternative ideas. It also underscores the importance of fostering a culture of skepticism and open inquiry, where dissenting voices are valued rather than silenced.

The Role of Science and Philosophy in Overcoming Fallacies

The fallacies of the human mind highlight the limitations of individual cognition, but they also demonstrate the power of collective, systematic inquiry to overcome these limitations. Science, with its emphasis on empirical evidence and falsifiability, provides a framework for correcting errors and refining knowledge over time. Philosophy, meanwhile, offers tools for examining the assumptions and biases that underlie our thinking.

By acknowledging and addressing these fallacies, we can cultivate a mindset of humility and curiosity—qualities that are essential for the pursuit of truth. The next chapter will explore how science and philosophy shape our understanding of existence, weaving together the threads of perception, thought, and reality to form a richer picture of the human experience.

The Fallacy of Anthropocentrism: The Human Lens on the Universe

One of the most pervasive and subtle fallacies is anthropocentrism—the tendency to view the universe through a human-centered lens. This fallacy stems from the natural inclination to interpret reality in terms of human experience, assuming that our place in the cosmos is special or central. Anthropocentrism has shaped everything from ancient cosmologies to modern scientific debates, influencing how we perceive our relationship with the natural world.

In antiquity, anthropocentrism was embodied in the **geocentric model** of the universe, which placed Earth at the center of creation. This model, championed by figures like Ptolemy and deeply entwined with religious doctrine, reflected the belief that humanity occupied a privileged position in the cosmos. The heliocentric revolution initiated by Copernicus, and later supported by Galileo and Kepler, was a profound challenge to this worldview. By displacing Earth from the center of the universe, it forced humanity to confront the humbling reality that our planet is just one among many celestial bodies orbiting an average star in a vast and indifferent cosmos.

Even in the modern era, anthropocentrism persists in subtler forms. Consider the search for extraterrestrial life. Many scientists have focused their efforts on identifying Earth-like planets, assuming that life elsewhere must resemble life as we know it. This assumption, while pragmatic, reflects a bias rooted in human experience. What if life in other parts of the universe operates on entirely different principles, using chemical elements or energy sources that are alien to our biology? The anthropocentric fallacy limits our ability to conceive of possibilities beyond the familiar, underscoring the importance of cultivating intellectual humility in scientific inquiry.

Philosophical Implications:
Anthropocentrism raises profound questions about the nature of knowledge and perception. To what extent are our scientific models shaped by human biases? Can we ever fully escape the constraints of our cognitive frameworks to perceive the universe as it truly is? Philosophers like Thomas Nagel have explored these questions, arguing that certain aspects of reality may be forever inaccessible to human understanding. This perspective challenges the idea that science can achieve a "God's-eye view" of the universe, suggesting instead that our knowledge is inevitably shaped by our limited, human vantage point.

The Fallacy of Presentism: The Illusion of Now

Another common cognitive fallacy is presentism—the tendency to view history and the future through the lens of the present, assuming that current beliefs, values, and knowledge are superior to those of the past and predictive of the future. Presentism manifests in the assumption that the scientific and philosophical frameworks of our time are the pinnacle of human achievement, overlooking the provisional nature of knowledge and the potential for future paradigm shifts.

Example: The Limitations of Current Theories

Consider the Standard Model of particle physics, one of the most successful scientific theories ever developed. It accurately describes the behavior of fundamental particles and forces, except for gravity. Despite its successes, the Standard Model cannot account for dark matter, dark energy, or the quantum nature of gravity—phenomena that constitute the majority of the universe. While physicists acknowledge these gaps, presentism can lead to overconfidence in the completeness of the model, obscuring the possibility that future discoveries may render it as incomplete as Newtonian mechanics in the age of relativity.

Presentism is also evident in ethical and philosophical debates. For instance, many modern societies view their moral frameworks as the culmination of centuries of progress, yet history demonstrates that ethical systems are shaped by cultural and historical contexts. Practices once considered morally acceptable, such as slavery or colonialism, are now universally condemned, while contemporary issues like artificial intelligence and genetic engineering pose new ethical challenges that future generations may judge harshly.

Philosophical Implications:
Presentism highlights the tension between progress and humility. While science and philosophy build on the achievements of the past, they must also remain open to the possibility that current frameworks are flawed or incomplete. This requires a balance between confidence in our knowledge and a recognition of its limitations—a principle exemplified by the philosophy of Karl Popper, who emphasized the importance of falsifiability and the iterative nature of scientific progress.

The Availability Heuristic: The Trap of Familiar Evidence

Another critical fallacy that distorts human reasoning is the **availability heuristic**, a cognitive shortcut in which people judge the likelihood or importance of an event based on how easily examples come to mind. This fallacy leads us to overestimate the frequency of dramatic, highly publicized events, such as plane crashes or terrorist attacks, while underestimating more common but less sensational risks, such as car accidents or heart disease.

In science, the availability heuristic can skew the interpretation of evidence. Researchers may focus on well-documented phenomena while neglecting less visible but equally important factors. For example, early research on climate change emphasized rising temperatures because these effects were immediately observable, while the impact of ocean acidification—an equally critical consequence of increased carbon dioxide levels—received less attention due to its less dramatic presentation.

Philosophical Implications:
The availability heuristic underscores the importance of systematic methods for evaluating evidence. Unlike intuitive reasoning, which is prone to bias, scientific inquiry relies on tools such as statistics, peer review, and replication to ensure that judgments are based on the full range of evidence rather than the most salient examples. This methodological rigor reflects

the epistemological principle that truth is often obscured by human perception, requiring deliberate effort to uncover.

The Sunk Cost Fallacy: Clinging to Past Investments

The sunk cost fallacy occurs when individuals or institutions continue to invest in a failing endeavor because of the resources they have already committed. This fallacy is driven by a reluctance to admit error and a desire to justify past decisions, even when abandoning the effort would be more rational.

Example: The Persistence of Outdated Theories
In the history of science, the sunk cost fallacy has contributed to the persistence of outdated theories. For instance, proponents of the caloric theory of heat, which posited that heat was a fluid-like substance, resisted abandoning the theory even as evidence for thermodynamics mounted. Their commitment to caloric theory was not merely intellectual but also emotional and professional, reflecting the difficulty of letting go of ideas in which they had invested years of research.

Broader Implications:
The sunk cost fallacy is not limited to science; it permeates politics, business, and personal decision-making. For example, governments may continue funding ineffective programs because of the resources already spent, while individuals may remain in unhealthy relationships or unfulfilling careers for similar reasons. Recognizing this fallacy requires the courage to reassess priorities and the humility to admit past mistakes—a principle that is central to both scientific and philosophical inquiry.

The Intersection of Science and Fallibility

These cognitive fallacies—anthropocentrism, presentism, the availability heuristic, and the sunk cost fallacy—illustrate the profound challenges of navigating a world where human perception and reasoning are inherently flawed. Yet they also underscore the resilience of science and philosophy in overcoming these limitations. By identifying and addressing fallacies, we can refine our methods, broaden our perspectives, and approach truth with greater clarity and humility.

Science, with its emphasis on falsifiability, peer review, and empirical testing, provides a framework for minimizing the influence of fallacies. Philosophy, meanwhile, offers tools for examining the assumptions and biases that underlie human thought. Together, these disciplines remind us that fallibility is not a weakness but an opportunity for growth—a reminder that the pursuit of truth is a journey, not a destination.

The Illusion of Causality: Mistaking Correlation for Cause

One of the most pervasive cognitive fallacies in human reasoning is the illusion of causality—the tendency to assume that when two events occur together, one must have caused the other. This fallacy arises from the human brain's natural inclination to seek patterns and explanations, a trait that has evolutionary advantages but often leads to erroneous conclusions. In science, philosophy, and everyday life, the illusion of causality can distort understanding and impede progress.

Example: The "Rain and Roosters" Problem

Consider a hypothetical scenario in which villagers observe that rain frequently begins shortly after a rooster crows. Over time, they conclude that the rooster's crowing causes the rain. This error stems from the fact that the two events are correlated—they often occur in close temporal proximity—but are not causally connected. The rooster's crow and the onset of rain are both independent phenomena, linked only by coincidence or underlying factors such as the time of day.

In the history of science, similar errors have often occurred. For centuries, physicians believed in the theory of "miasma"—the idea that diseases were caused by bad air—because of the observed correlation between foul odors and outbreaks of illness. It was only through the work of figures like Louis Pasteur and Robert Koch that the true causes of disease—microorganisms—were identified, overturning the flawed assumptions of the miasma theory.

Philosophical Implications:

The illusion of causality highlights the difficulty of distinguishing between correlation and causation, a challenge that lies at the heart of scientific inquiry. The philosopher David Hume famously argued that causation is not directly observable; we infer it from patterns of regularity. Hume's skepticism reminds us that even the most intuitive causal relationships must be rigorously tested. Modern science addresses this challenge through controlled experiments, statistical analysis, and the principle of falsifiability, ensuring that claims of causation are supported by evidence rather than assumption.

The Dunning-Kruger Effect: The Pitfalls of Overestimating Competence

Another significant cognitive fallacy is the **Dunning-Kruger effect**, a psychological phenomenon in which individuals with low expertise in a domain overestimate their knowledge and abilities. Conversely, those with high expertise often underestimate their competence, as they are more aware of the complexities and limitations of their field. This fallacy not only affects individuals but can also shape societal discourse, leading to the proliferation of misinformation and the undervaluing of genuine expertise.

Example: The Era of Misinformation

The rise of the internet and social media has amplified the Dunning-Kruger effect on a global scale. Platforms that prioritize visibility over accuracy allow individuals with limited understanding to present themselves as authorities, while true experts are often drowned out by louder, less informed voices. This phenomenon is particularly evident in debates about climate change, public health, and vaccination. For example, during the COVID-19 pandemic, individuals with little to no scientific training spread misinformation about vaccines, challenging the findings of decades of immunological research. This misinformation, rooted in the overconfidence of the uninformed, undermined public trust in science and hindered efforts to control the pandemic.

Philosophical Implications:
The Dunning-Kruger effect raises important questions about epistemology and the nature of knowledge. How can society distinguish between genuine expertise and false confidence? What responsibilities do experts have in communicating their findings to the public? And how can educational systems cultivate intellectual humility while empowering individuals to think critically? Addressing these questions requires a multifaceted approach, combining better public education, stronger science communication, and the promotion of evidence-based decision-making.

The Appeal to Authority: When Expertise Becomes Dogma

While the undervaluing of expertise is a major problem, its opposite—the **appeal to authority**—is an equally significant fallacy. This fallacy occurs when individuals accept a claim as true simply because it comes from a perceived authority figure, without critically examining the evidence. The appeal to authority can lead to intellectual stagnation, as ideas are accepted based on status rather than merit.

Example: Aristotle and the Long Shadow of Classical Authority
For centuries, the works of Aristotle were treated as the ultimate source of knowledge in many fields, from physics to biology. His ideas about the natural world, such as the belief that heavier objects fall faster than lighter ones, were accepted without question. This reverence for Aristotle's authority delayed the progress of science, as alternative ideas were often dismissed outright. It was only with the experiments of Galileo Galilei in the 16th century that the Aristotelian worldview began to crumble, paving the way for the scientific revolution.

Philosophical Implications:
The appeal to authority highlights the tension between respect for expertise and the need for skepticism. While expertise is valuable, it must never become a substitute for critical inquiry. This principle is central to the philosophy of science, which emphasizes the provisional nature of knowledge and the importance of questioning even the most established ideas. As Carl Sagan famously said, "Science is a way of thinking much more than it is a body of knowledge."

The Narrative Fallacy: The Search for Coherence in Chaos

Humans are natural storytellers, and we instinctively seek to organize complex events into coherent narratives. This tendency, while useful for understanding the world, often leads to the **narrative fallacy**—the oversimplification of reality to fit a preferred story. In science, philosophy, and history, the narrative fallacy can obscure complexity and perpetuate myths.

Example: The Myth of the "Lone Genius"
One common narrative fallacy in the history of science is the "lone genius" myth—the idea that groundbreaking discoveries are the work of solitary individuals. Figures like Isaac Newton, Albert Einstein, and Marie Curie are often depicted as isolated geniuses, toiling alone to uncover the secrets of the universe. While these individuals made extraordinary

contributions, their work was deeply embedded in broader networks of collaboration, mentorship, and prior research. For example, Einstein's theory of relativity built on the insights of Hendrik Lorentz, Henri Poincaré, and James Clerk Maxwell, among others.

Philosophical Implications:
The narrative fallacy reminds us that truth is often messier than the stories we tell about it. It challenges us to embrace complexity and resist the temptation to reduce multifaceted phenomena to simplistic explanations. In philosophy, this principle is reflected in the idea of "anti-reductionism," which emphasizes the interconnectedness of knowledge and the need to consider multiple perspectives.

The Status Quo Bias: Clinging to the Familiar

The **status quo bias** is the tendency to prefer existing conditions over change, even when change would lead to better outcomes. This fallacy arises from a combination of fear, inertia, and loss aversion—the psychological preference for avoiding losses over achieving gains.

Example: Resistance to Renewable Energy
In the context of climate change, the status quo bias is a major barrier to the adoption of renewable energy technologies. Despite the clear benefits of transitioning to solar, wind, and other sustainable energy sources, many governments and industries remain invested in fossil fuels. This resistance is often justified by concerns about economic disruption or the cost of infrastructure changes, even though the long-term costs of inaction are far greater.

Philosophical Implications:
The status quo bias underscores the ethical dimension of decision-making. How do we balance the risks of change against the dangers of complacency? In philosophy, this tension is reflected in debates about conservatism and progressivism, as well as the ethics of precautionary principles. Overcoming the status quo bias requires not only rational analysis but also the courage to envision and pursue alternative futures.

The Availability Cascade: How Repetition Shapes Belief

One of the more insidious cognitive fallacies is the **availability cascade**, a process by which repeated exposure to a claim, regardless of its accuracy, increases its perceived validity. This phenomenon occurs because the human brain tends to equate familiarity with truth—a shortcut that often leads to the spread of misinformation and the entrenchment of false beliefs. The availability cascade is not only a psychological curiosity but also a profound obstacle to rational thought in science, philosophy, and society.

Example: The Myth of "Humans Use Only 10% of Their Brain"
The pervasive myth that humans use only 10% of their brain power exemplifies the availability cascade. This claim, despite being thoroughly debunked by neuroscience, persists in popular culture due to its frequent repetition in movies, self-help books, and media. Studies using functional MRI (fMRI) and PET scans have shown that virtually every part of the brain

has a function, even during simple tasks like resting. Yet the persistence of this myth demonstrates how repetition can override evidence, embedding itself in public consciousness.

The availability cascade is particularly dangerous in science, where it can lead to the premature acceptance of theories or the dismissal of alternatives. For instance, early in the 20th century, the caloric theory of heat persisted long after experimental evidence had undermined it, simply because it was entrenched in scientific discourse. Similarly, public debates on topics like climate change or vaccine safety are often shaped more by the repetition of claims in media than by the weight of evidence.

Philosophical Implications:
The availability cascade raises important questions about the nature of belief and its relationship to truth. How do we distinguish between ideas that are valid and those that merely feel familiar? The philosopher Friedrich Nietzsche once observed that "convictions are more dangerous enemies of truth than lies," a statement that resonates in the context of availability cascades. In an age of information overload, critical thinking and skepticism are more essential than ever, reminding us that familiarity should never be mistaken for evidence.

The Halo Effect: How First Impressions Cloud Judgment

The **halo effect** is a cognitive bias in which our overall impression of a person, group, or idea influences how we evaluate their specific traits. This bias often occurs unconsciously, leading us to overgeneralize from a single positive or negative attribute. While the halo effect is most commonly discussed in the context of personal relationships or marketing, it has profound implications for science, education, and public discourse.

Example: Reverence for "Genius" Figures in Science
In the history of science, the halo effect can be seen in the reverence afforded to certain "genius" figures, such as Isaac Newton, Albert Einstein, or Charles Darwin. While their contributions to science are undeniable, the halo effect often leads to an uncritical acceptance of all their ideas, even those that have been proven incorrect or are outside their areas of expertise. For example, Newton's later work on alchemy and biblical prophecy, while historically interesting, is far less rigorous than his contributions to physics and mathematics. Yet the halo surrounding Newton's name often shields these less credible pursuits from the scrutiny they deserve.

The same bias can also operate in reverse, where a single perceived flaw in a scientist's character or work leads to the dismissal of their valid contributions. For instance, the personal eccentricities of figures like Nikola Tesla or Paul Dirac have sometimes overshadowed their groundbreaking work in electromagnetism and quantum mechanics, respectively. This dual nature of the halo effect illustrates how biases rooted in personality and perception can distort the evaluation of ideas.

Philosophical Implications:
The halo effect challenges the principle of objectivity, reminding us that human judgment is often colored by irrelevant factors. In philosophy, this bias connects to broader questions about authority and credibility. Should ideas stand or fall on their own merit, independent of their source? The philosopher Immanuel Kant's emphasis on the "autonomy of reason"

suggests that truth must be evaluated independently of its author, a principle that remains as relevant today as it was in the Enlightenment.

The Endowment Effect: Overvaluing What We Own

The **endowment effect** is a cognitive bias in which people ascribe greater value to things simply because they own them. This bias has been extensively studied in behavioral economics, where it explains phenomena like the reluctance to sell possessions at market price, even when doing so would be rational. However, the endowment effect extends beyond material goods, influencing how people evaluate ideas, beliefs, and intellectual frameworks.

Example: Clinging to Outdated Scientific Theories
In science, the endowment effect can manifest as a reluctance to abandon outdated theories or paradigms, even in the face of contradictory evidence. For example, the geocentric model of the universe persisted for centuries because it was deeply embedded in both scientific and cultural frameworks. Scholars and institutions had invested so heavily in this model—through texts, teachings, and theological interpretations—that letting it go required not only intellectual shifts but also emotional and institutional reckonings.

Even today, the endowment effect shapes debates in fields like medicine, where long-held practices such as the routine use of antibiotics or the reliance on body mass index (BMI) as a measure of health are increasingly questioned. The resistance to change often stems not from evidence but from the psychological attachment to familiar ideas and practices.

Philosophical Implications:
The endowment effect highlights the role of attachment in shaping belief systems. Why do humans cling to ideas, even when they no longer serve us? This question resonates with the work of existentialist philosophers like Søren Kierkegaard, who argued that human beings often resist change because it threatens their sense of identity and security. Overcoming the endowment effect requires intellectual courage—the willingness to let go of cherished ideas in the pursuit of deeper truths.

The Curse of Knowledge: When Expertise Becomes a Barrier

The **curse of knowledge** is a cognitive bias in which individuals who possess expertise or familiarity with a topic struggle to imagine what it is like to lack that knowledge. This bias can create significant barriers to communication, particularly in education and public discourse, where experts must convey complex ideas to non-specialist audiences.

Example: The Challenges of Science Communication
The curse of knowledge is a major challenge in fields like physics and climate science, where experts often assume that key concepts—such as quantum superposition or the greenhouse effect—are self-evident. As a result, their explanations may fail to address the gaps in understanding that exist for lay audiences. This communication gap can lead to public misunderstandings, skepticism, or even rejection of scientific findings.

For instance, the idea that "climate change is caused by greenhouse gases trapping heat" is often presented without sufficient explanation of the underlying mechanisms. While this shorthand may be clear to scientists, it can seem abstract or unconvincing to non-specialists, contributing to confusion and resistance. Overcoming the curse of knowledge requires not only clarity but also empathy—the ability to see the world through the eyes of someone encountering the subject for the first time.

Philosophical Implications:
The curse of knowledge raises questions about the nature of expertise and the responsibility of those who possess it. How can experts bridge the gap between specialized knowledge and public understanding? And how do we ensure that the pursuit of truth remains accessible to all, rather than confined to an elite few? These questions resonate with the democratic ideals of the Enlightenment, which sought to make knowledge a public good rather than a privilege.

The Zeigarnik Effect: The Power of Unfinished Business

The **Zeigarnik effect** describes the human tendency to remember incomplete tasks or unresolved issues more vividly than completed ones. While this phenomenon has practical benefits, such as motivating individuals to finish what they start, it can also lead to fixation and anxiety, particularly in intellectual pursuits.

Example: Unanswered Questions in Science
The Zeigarnik effect is evident in the enduring fascination with unresolved scientific questions, such as the nature of dark matter, the origin of consciousness, or the unification of quantum mechanics and general relativity. These "unfinished" problems occupy a central place in scientific discourse, driving research and speculation. However, the same effect can also lead to obsessive focus on questions that may be unanswerable, such as the ultimate purpose of the universe or the existence of a multiverse.

Philosophical Implications:
The Zeigarnik effect underscores the human need for closure and its role in shaping intellectual inquiry. It raises profound questions about the limits of knowledge: Are some questions inherently unanswerable, and if so, how should we approach them? The philosopher Ludwig Wittgenstein famously argued that some problems arise not from reality itself but from the limitations of language and thought. This perspective invites us to consider whether the drive for resolution is always justified—or whether some mysteries are best left unresolved.

The Optimism Bias: Overestimating the Best Possible Outcomes

One of the most deeply ingrained cognitive biases is the **optimism bias**, the tendency to overestimate the likelihood of positive outcomes while underestimating risks and negative possibilities. This bias is not merely an occasional mental misstep—it is hardwired into the human brain, likely as an evolutionary mechanism to encourage perseverance in the face of adversity. However, while optimism bias can foster resilience and hope, it often leads to poor decision-making, both at the individual level and within scientific and philosophical contexts.

Example: Overconfidence in Technological Solutions
The optimism bias is frequently seen in debates about global challenges such as climate change, where there is a widespread assumption that future technological innovations will resolve environmental crises. While technology has undoubtedly played a transformative role in addressing past problems, this assumption can lead to complacency, delaying critical action in the present. For instance, reliance on unproven technologies such as carbon capture or geoengineering, while promising, ignores the complexities and risks associated with deploying these solutions at scale.

Similarly, in the early days of nuclear power, optimism bias led to grand visions of a future powered entirely by clean, limitless energy. Yet this optimism often failed to account for challenges such as radioactive waste, reactor safety, and the geopolitical risks of nuclear proliferation. While nuclear power remains a vital part of the global energy mix, its history underscores the dangers of assuming that innovation will always deliver unproblematic solutions.

Philosophical Implications:
The optimism bias raises questions about the tension between hope and realism in human thought. Should we temper our aspirations with skepticism, or does optimism itself drive the innovation and perseverance necessary for progress? Philosophers such as Friedrich Nietzsche have argued that a "will to power"—an embrace of ambition and optimism—fuels human achievement, while others, like the Stoics, advocate for cautious realism. The optimism bias thus forces us to reflect on the balance between dreaming of better futures and confronting present realities.

The "Just World" Fallacy: The Comfort of Moral Order

Another pervasive fallacy is the **just world fallacy**, the belief that the world is inherently fair and that people get what they deserve. This cognitive bias arises from a psychological need to impose order on a chaotic universe, creating a sense of moral equilibrium where virtue is rewarded, and wrongdoing is punished. While this belief can provide comfort and motivation, it often distorts reality, leading to victim-blaming, oversimplification of complex systems, and resistance to acknowledging systemic injustices.

Example: Misinterpretations in Economics and Social Systems
The just world fallacy is frequently seen in discussions about wealth and poverty. Many people, consciously or unconsciously, assume that the wealthy have earned their success through hard work and intelligence, while the poor are responsible for their own struggles. While individual effort and choices certainly play a role, this oversimplification ignores the structural factors—such as access to education, systemic inequality, and historical injustices—that shape economic outcomes.

In science, the just world fallacy can lead to distorted interpretations of evolutionary theory, such as the misapplication of "survival of the fittest" to justify social hierarchies or competition. These interpretations, often associated with Social Darwinism, reflect a flawed attempt to impose moral order on natural processes, overlooking the complexities of cooperation, mutualism, and contingency in evolution.

Philosophical Implications:
The just world fallacy challenges us to confront the discomforting reality that the universe is not inherently fair. In philosophy, this tension is explored in existentialist thought, particularly in the work of Albert Camus and his concept of the "absurd." Camus argued that human beings seek meaning and justice in a world that offers neither, and that the task of philosophy is to navigate this contradiction without succumbing to false comforts. The just world fallacy thus reflects a deeper struggle to reconcile human values with an indifferent cosmos.

The Anchoring Effect: How Initial Information Skews Judgment

The **anchoring effect** is a cognitive bias in which people rely too heavily on the first piece of information they receive when making decisions, even when subsequent evidence contradicts it. This "anchor" can distort judgment, leading to flawed conclusions in fields ranging from economics to medicine to policy-making.

Example: Anchoring in Scientific Paradigms
The anchoring effect is particularly relevant in the development and evaluation of scientific theories. For example, early models of the atom, such as the "plum pudding" model proposed by J.J. Thomson, served as anchors that influenced subsequent research. Even as evidence mounted against this model—such as Ernest Rutherford's gold foil experiment, which revealed the nucleus—its initial framework shaped how scientists interpreted new data.

Similarly, in modern medicine, anchoring can occur when doctors form an initial diagnosis based on limited information, such as a patient's description of symptoms. This initial impression can bias subsequent tests and evaluations, potentially leading to misdiagnosis. Overcoming the anchoring effect requires deliberate effort to remain open to alternative interpretations and to continuously update beliefs based on new evidence.

Philosophical Implications:
The anchoring effect highlights the difficulty of achieving true intellectual flexibility. How do we break free from the influence of initial assumptions? This question resonates with the work of philosophers like John Stuart Mill, who emphasized the importance of continuously exposing oneself to diverse perspectives to counteract the narrowing effects of dogmatism. In a broader sense, the anchoring effect illustrates the tension between stability and adaptability in human thought—a balance that is critical to both science and philosophy.

The Backfire Effect: When Evidence Strengthens False Beliefs

One of the most counterintuitive cognitive biases is the **backfire effect**, in which presenting evidence that contradicts someone's beliefs can actually strengthen those beliefs. This paradox arises because people often interpret contradictory evidence as a threat to their identity, prompting them to double down on their original views to protect their sense of self.

Example: Resistance to Scientific Consensus
The backfire effect is particularly evident in debates about issues such as climate change,

vaccination, and evolution. For instance, individuals who reject the scientific consensus on climate change often respond to evidence by seeking out information that reinforces their skepticism, such as pseudoscientific articles or conspiracy theories. This reaction is not merely a rejection of facts but a defense of identity, as beliefs about climate change are often tied to broader ideological or cultural affiliations.

The backfire effect is also observed in political discourse, where attempts to correct misinformation frequently fail. Studies have shown that fact-checking campaigns, while valuable, sometimes have the unintended consequence of entrenching false narratives among those already predisposed to distrust the source of the correction.

Philosophical Implications:
The backfire effect raises profound questions about the relationship between belief and identity. Can reason ever overcome deeply held biases, or are some beliefs impervious to evidence? Philosophers such as Hannah Arendt have explored the interplay between truth, power, and persuasion, arguing that truth alone is insufficient to change minds without addressing the emotional and social contexts in which beliefs are embedded. The backfire effect underscores the importance of empathy and dialogue in bridging divides, reminding us that knowledge is not just a matter of facts but also of trust and connection.

The Planning Fallacy: Underestimating Complexity and Time

The **planning fallacy** describes the human tendency to underestimate the time, resources, and effort required to complete tasks. This bias affects individual projects, organizational planning, and even large-scale scientific endeavors, often leading to delays, cost overruns, and unmet expectations.

Example: The Human Genome Project
The Human Genome Project, launched in 1990, aimed to sequence the entire human genome within 15 years. While the project ultimately succeeded, it faced significant delays and technical challenges, highlighting the planning fallacy's impact on even the most meticulously organized scientific efforts. The complexity of mapping billions of base pairs, combined with unforeseen hurdles in data analysis and technology development, required continuous adjustments to the original timeline and budget.

The planning fallacy is also evident in space exploration, where ambitious projects like the James Webb Space Telescope or human missions to Mars have repeatedly faced delays due to underestimated technical challenges and funding shortfalls. These examples underscore the difficulty of anticipating all variables in complex systems, reminding us that the path from vision to reality is rarely straightforward.

Philosophical Implications:
The planning fallacy invites reflection on the limits of human foresight. Can we ever fully account for complexity, or are we doomed to perpetual underestimation? This question resonates with the work of systems theorists, who emphasize the interconnectedness and unpredictability of large-scale endeavors. The planning fallacy serves as a humbling reminder of the gap between intention and execution, challenging both our optimism and our hubris.

Conclusion: The Mind as Both Strength and Limitation

The fallacies of the human mind—from the illusion of causality to the status quo bias—highlight the paradox of human cognition. Our ability to reason, imagine, and learn is what allows us to explore the mysteries of the universe, yet these same abilities are constrained by biases, shortcuts, and errors. Recognizing these limitations is not an admission of defeat but a call to action. By understanding our fallacies, we can design systems and methods—such as the scientific method—that help us overcome them, inching closer to truth.

As we move into the next chapter, we will explore how science and philosophy converge to address the ultimate questions of existence, reconciling the fallibility of the human mind with the pursuit of meaning and truth.

Chapter 8: Scientia, Philosophia, et Existentia

(Science, Philosophy, and Existence)

Introduction: The Human Quest for Meaning

From the dawn of humanity, people have sought to understand their place in the universe. Science and philosophy, while distinct in their methods, have been united in this quest, asking questions about existence, reality, and purpose. Science reveals the intricate mechanisms of the natural world, from the smallest particles to the vastness of the cosmos. Philosophy, meanwhile, explores the meaning of these discoveries, probing questions that lie beyond empirical measurement: What is the nature of being? Why does the universe exist at all? How should we live in a world that appears indifferent to our presence?

This chapter delves into the interplay between science and philosophy as they seek to illuminate the mysteries of existence. By examining the boundaries of scientific knowledge, the insights of existentialist thinkers, and the role of perception and consciousness, we will explore how humanity navigates the tension between a universe governed by objective laws and the deeply subjective experience of being alive.

The Scientific Framework: Explaining the Universe

Science is one of humanity's greatest tools for understanding existence. It seeks to explain the "how" of the universe: How do galaxies form? How do atoms interact? How does consciousness emerge from neural activity? From the Copernican revolution to the discovery of DNA, science has revealed that the universe operates according to patterns and laws that can be measured, predicted, and modeled. Yet these explanations, no matter how precise, often provoke deeper philosophical questions about the "why" of existence.

Example: The Origins of the Universe
Consider the scientific explanation for the origin of the universe: the Big Bang theory. According to this model, the universe began as a singularity approximately 13.8 billion years ago, expanding and cooling to form the galaxies, stars, and planets we observe today. This theory is supported by extensive evidence, such as the cosmic microwave background radiation and the redshift of distant galaxies.

Yet the Big Bang theory raises profound philosophical questions. What caused the singularity to exist? What, if anything, came before it? Is the universe finite or infinite? While science provides models to describe the universe's evolution, it often leaves these metaphysical questions unanswered, inviting philosophy to grapple with the deeper implications of our cosmic origins.

Existentialism: Finding Meaning in a Scientific World

The rise of modern science has profoundly shaped how humans think about existence. As scientific discoveries have displaced traditional, anthropocentric views of the cosmos, they have often challenged long-held beliefs about humanity's significance. In response to this shift, existentialist philosophy emerged as a framework for understanding the human condition in a universe that seems indifferent to individual lives and desires.

The Absurd: Navigating a Meaningless Universe
Existentialists such as Albert Camus and Jean-Paul Sartre grappled with the concept of the "absurd"—the conflict between humanity's desire for meaning and the apparent

meaninglessness of the universe. For Camus, the absurd was not a problem to be solved but a condition to be embraced. In *The Myth of Sisyphus*, he likens the human struggle for meaning to the plight of Sisyphus, a figure from Greek mythology condemned to roll a boulder up a hill for eternity. While Sisyphus's task is futile, Camus argues that his defiance in continuing to push the boulder gives his existence a kind of meaning.

This perspective resonates with scientific discoveries that challenge human centrality. For example, the realization that Earth is a tiny planet orbiting an average star in an unremarkable galaxy underscores the existentialist idea that meaning is not inherent in the universe but must be created by individuals. Science and existentialism thus converge in their recognition of the vastness and indifference of the cosmos, offering complementary tools for navigating this realization.

The Role of Perception: Science, Consciousness, and Subjectivity

While science seeks objective truths, existence is experienced subjectively, through the lens of human perception and consciousness. This duality creates a tension between the external reality described by science and the internal reality of lived experience.

Example: The Hard Problem of Consciousness

The "hard problem of consciousness," a term coined by philosopher David Chalmers, exemplifies this tension. While neuroscience can explain how brain activity correlates with mental states, it struggles to explain why these processes give rise to subjective experience—the rich, inner life that includes thoughts, emotions, and sensations. Why does the firing of neurons produce the feeling of pain or the perception of color?

This question lies at the intersection of science and philosophy, challenging the reductionist assumption that all phenomena can be explained in purely physical terms. For thinkers like Roger Penrose, consciousness may involve quantum processes that go beyond the scope of classical computation. Others, such as existentialists, argue that consciousness is defined not by its mechanisms but by its capacity to grapple with questions of freedom, responsibility, and meaning.

Philosophical Implications:

The hard problem of consciousness underscores the limits of scientific explanation, reminding us that some aspects of existence may remain inherently mysterious. At the same time, it highlights the importance of subjectivity in shaping human understanding. While science offers a framework for describing the universe, consciousness gives meaning to this framework, transforming abstract facts into lived experience.

Freedom and Determinism: Are We Free in a Scientific Universe?

One of the most profound questions about existence is whether human beings possess free will or are governed entirely by deterministic laws. This debate lies at the heart of both scientific inquiry and philosophical reflection.

Scientific Determinism:

The rise of classical physics in the 17th century, particularly the work of Isaac Newton, suggested that the universe operates as a vast, deterministic machine. If the motion of every particle is governed by predictable laws, then, in principle, the future could be determined entirely by the present. This view, often called "Laplacean determinism," raises troubling questions about human freedom: If all actions are the result of prior causes, can we truly be said to choose our actions?

Quantum Mechanics and Indeterminacy:

The advent of quantum mechanics in the 20th century complicated this deterministic picture. At the quantum level, particles do not behave predictably but probabilistically, governed by the uncertainties described in Heisenberg's uncertainty principle. Some philosophers and scientists have argued that this indeterminacy reopens the possibility of free will. However, others contend that randomness is no better a foundation for free will than determinism, as it replaces causation with unpredictability.

Existential Freedom:

Existentialist philosophers, such as Jean-Paul Sartre, approach the question of freedom from a different angle. For Sartre, human freedom is not about escaping causality or randomness but about the capacity to make choices and take responsibility for them. In Sartre's view, even in a deterministic universe, individuals are "condemned to be free" because they must continually define themselves through their actions. This existential perspective offers a way to reconcile the scientific description of the universe with the lived reality of choice and agency.

The Limits of Knowledge: Embracing Mystery

Both science and philosophy strive to understand existence, yet both are bound by limits. Science, for all its rigor, cannot fully explain why there is something rather than nothing, nor can it address questions of ultimate purpose. Philosophy, while capable of grappling with these questions, often offers answers that are deeply subjective and contingent on individual perspectives.

Example: Gödel's Incompleteness Theorems

The work of mathematician Kurt Gödel illustrates the inherent limits of knowledge. Gödel's incompleteness theorems show that in any sufficiently complex system of logic, there are truths that cannot be proven within the system itself. This insight, while grounded in mathematics, has profound philosophical implications, suggesting that some aspects of reality may lie beyond the reach of human comprehension.

In a similar vein, the principles of epistemology remind us that knowledge is always provisional, subject to revision in light of new evidence. This humility is central to both science and philosophy, encouraging us to embrace uncertainty as an integral part of the human condition.

Existence as a Shared Journey

Ultimately, the intersection of science, philosophy, and existence reflects the complexity of being human. Science offers tools for exploring the external world, while philosophy provides frameworks for understanding the internal world of thought, emotion, and meaning. Together, they remind us that existence is not a problem to be solved but a journey to be experienced—a constant interplay between the known and the unknown, the objective and the subjective, the finite and the infinite.

The Greek Era and the Quest for Knowledge

The ancient Greeks laid the foundation for Western thought, initiating a transformative era of intellectual inquiry that bridged mythology and reason. The Greek quest for knowledge was rooted in an insatiable curiosity about the natural world and humanity's place within it. Unlike earlier civilizations, where knowledge often served religious or practical purposes, the Greeks developed a systematic approach to understanding existence, characterized by logic, empirical observation, and the pursuit of universal principles. This intellectual revolution found expression in diverse fields, including **astronomy**, **medicine**, and **geometry**, where figures such as **Pythagoras**, **Hippocrates**, **Aristotle**, and **Euclid**made lasting contributions. Their achievements not only advanced specific disciplines but also shaped the philosophical and methodological frameworks that continue to underpin modern science.

Astronomy: Charting the Cosmos

Greek astronomy was deeply intertwined with philosophy, reflecting a desire to uncover the fundamental principles governing the heavens. While earlier civilizations, such as the Babylonians and Egyptians, had developed detailed astronomical observations for calendrical and navigational purposes, the Greeks sought to explain celestial phenomena through reason and geometry. This shift from description to explanation marked a turning point in the history of science.

Thales and the Birth of Rational Astronomy
The pre-Socratic philosopher **Thales of Miletus** (circa 624–546 BCE) is often credited as one of the first thinkers to propose natural explanations for cosmic events. Rejecting mythological accounts, Thales argued that celestial bodies were material entities governed by natural laws. He famously predicted a solar eclipse, demonstrating that human reason could anticipate seemingly divine occurrences. Though limited by the observational tools of his time, Thales' emphasis on rationality laid the groundwork for the systematic study of astronomy.

Ptolemy and the Geocentric Model
Greek astronomy reached its zenith with **Claudius Ptolemy** (circa 100–170 CE), whose *Almagest* became the definitive astronomical text for over a millennium. Ptolemy synthesized earlier Greek and Babylonian observations into a comprehensive geocentric model, in which the Earth stood immobile at the center of the universe, surrounded by nested spheres carrying the planets, Sun, and stars. Though incorrect, this model was a triumph of mathematical ingenuity, using epicycles to account for the apparent retrograde motion of planets. Ptolemy's work reflected the Greek belief that the cosmos could be understood through mathematical precision, a legacy that would later inspire figures such as Copernicus and Kepler.

Aristarchus and the Heliocentric Hypothesis

Amid the dominance of the geocentric view, **Aristarchus of Samos** (circa 310–230 BCE) proposed a bold alternative: that the Sun, not the Earth, was the center of the universe. While his heliocentric model was largely ignored in antiquity, Aristarchus' insights anticipated the Copernican revolution by nearly 1,800 years. His work demonstrated the Greek willingness to entertain radical ideas, even when they challenged deeply held assumptions.

Medicine: The Science of Healing

Greek medicine marked a departure from supernatural explanations of disease, emphasizing natural causes and systematic observation. The Greeks' holistic approach to health integrated the physical, mental, and environmental aspects of well-being, establishing a tradition that would shape medical practice for centuries.

Hippocrates and the Ethical Foundations of Medicine

The figure most associated with Greek medicine is **Hippocrates of Kos** (circa 460–370 BCE), often called the "Father of Medicine." Hippocrates rejected the idea that diseases were punishments from the gods, arguing instead that they arose from imbalances in the body's four humors: blood, phlegm, yellow bile, and black bile. His *Corpus Hippocraticum*, a collection of medical texts, emphasized the importance of careful observation, diagnosis, and prognosis, laying the foundation for evidence-based medicine.

Hippocrates is also remembered for the **Hippocratic Oath**, which established ethical guidelines for medical practitioners. The oath's emphasis on patient care, confidentiality, and non-maleficence reflects the Greek belief that medicine was not merely a technical skill but a moral vocation. This ethical framework continues to influence modern medical practice, underscoring the enduring relevance of Greek thought.

Galen and the Anatomy of the Body

Building on Hippocratic principles, **Galen of Pergamon** (circa 129–216 CE) advanced the study of anatomy and physiology. Though limited by cultural taboos against human dissection, Galen conducted detailed studies on animals, particularly apes, to infer the structure and function of human organs. His findings, compiled in texts such as *On the Usefulness of the Parts of the Body*, remained authoritative for over a thousand years. Galen's integration of philosophy and medicine reflected the Greek belief in the interconnectedness of mind and body, a theme that resonates in contemporary approaches to healthcare.

Geometry: The Language of the Universe

For the Greeks, geometry was not merely a practical tool for measurement but a means of understanding the universe's fundamental structure. The precision and abstraction of geometric reasoning appealed to their philosophical desire for universal truths, making geometry a cornerstone of both mathematics and metaphysics.

Pythagoras and the Harmony of Numbers

The philosopher and mathematician **Pythagoras** (circa 570–495 BCE) regarded numbers as the essence of all reality. His famous theorem, which relates the sides of a right triangle, was part of a broader belief that mathematical relationships underlie the harmony of the cosmos. Pythagoras and his followers explored the connections between geometry, music, and astronomy, asserting that the universe itself was a manifestation of numerical order.

Pythagoras' emphasis on the metaphysical significance of mathematics influenced later thinkers, including Plato, who regarded geometry as a pathway to philosophical enlightenment. The inscription above the entrance to Plato's Academy—"Let no one ignorant of geometry enter"—reflects the centrality of mathematics in Greek intellectual life.

Euclid and the Foundations of Geometry

Euclid of Alexandria (circa 300 BCE) systematized the principles of geometry in his seminal work, *Elements*, which became one of the most influential texts in the history of mathematics. Comprising 13 books, *Elements* presented a rigorous deductive framework, starting from a small set of axioms and building a comprehensive system of geometric theorems. Euclid's method of logical proof set a standard for mathematical rigor that endures to this day.

Archimedes and Applied Geometry

While Euclid focused on theoretical geometry, **Archimedes of Syracuse** (circa 287–212 BCE) applied geometric principles to solve practical problems. Archimedes invented ingenious devices, such as water screws and siege engines, and made significant contributions to the understanding of volume, surface area, and the principles of leverage. His work demonstrated the interplay between abstract reasoning and practical innovation, a hallmark of Greek science.

The Legacy of Greek Inquiry

The Greek quest for knowledge, exemplified by their achievements in astronomy, medicine, and geometry, reflects a profound commitment to understanding the world through reason, observation, and systematic thought. Their intellectual traditions laid the groundwork for the scientific method, inspiring later civilizations in the Islamic world, Renaissance Europe, and beyond. Yet, Greek science was not merely a collection of techniques and discoveries; it was a philosophical endeavor, rooted in the belief that knowledge of the natural world was inseparable from the pursuit of truth and virtue.

This legacy reminds us that the human quest for knowledge is both universal and timeless, transcending cultural and historical boundaries. By studying the Greeks, we not only honor their achievements but also reconnect with the spirit of inquiry that defines our shared humanity.

The Islamic Golden Age and the Quest for Knowledge

The **Islamic Golden Age**, spanning roughly from the 8th to the 14th centuries, was a period of remarkable scientific, philosophical, and cultural achievements. During this era, scholars across the Islamic world sought to understand the mysteries of the universe, driven by a profound intellectual curiosity and a spirit of inquiry deeply rooted in the Qur'anic emphasis on knowledge (*ilm*). Far from being insular, the Islamic Golden Age was defined by its openness to ideas from other cultures. Muslim scholars actively translated, preserved, and expanded upon the works of the Greeks, Indians, Persians, and others, transforming ancient knowledge into new discoveries. This period laid the foundation for many fields of science, including astronomy, medicine, and geometry, which thrived under the patronage of caliphs and rulers who valued intellectual exploration as a form of worship and a path to understanding the divine.

Astronomy: Mapping the Heavens

Astronomy flourished during the Islamic Golden Age, driven by both practical needs—such as determining prayer times and the direction of the Kaaba in Mecca—and a desire to understand the cosmos. The Qur'an's references to the heavens and celestial bodies inspired scholars to study the skies, leading to significant advancements in observational techniques, mathematical models, and the design of instruments.

The House of Wisdom and the Translation Movement
The foundations of Islamic astronomy were laid at the **Bayt al-Hikmah** (*House of Wisdom*) in Baghdad, where scholars like **Al-Kindi, Hunayn ibn Ishaq**, and **Thabit ibn Qurra** translated Greek works, including Ptolemy's *Almagest*, into Arabic. However, these scholars did not merely preserve ancient knowledge; they expanded upon it. For example, **Al-Battani** (known in Latin as Albatenius), refined Ptolemy's models of planetary motion, introducing more accurate calculations of the orbits of the Sun and Moon. His work corrected many of Ptolemy's errors and influenced later astronomers, including Copernicus.

Al-Zarqali and the Andalousian Contributions
In Al-Andalus (modern-day Spain), **Al-Zarqali** (Arzachel) created highly accurate astrolabes and developed new models of planetary motion, including the "Toledo Tables," which were widely used in Europe for centuries. Al-Zarqali's work demonstrated a deep understanding of both practical and theoretical astronomy, and his ideas about the elliptical orbits of celestial bodies prefigured Kepler's later discoveries.

Ulugh Beg and Observatories
In the 15th century, **Ulugh Beg**, a Timurid ruler and astronomer, established a major observatory in Samarkand. His star catalog, which included precise measurements of over a thousand stars, was one of the most accurate of its time. Ulugh Beg's dedication to observational precision exemplified the Islamic world's commitment to empirical inquiry, bridging the gap between theoretical models and the physical universe.

Medicine: Healing the Body and Mind

Medicine was another area of significant advancement during the Islamic Golden Age, reflecting the era's holistic approach to health, which integrated physical, mental, and spiritual well-being. Building on the works of Hippocrates and Galen, Muslim physicians developed new treatments, surgical techniques, and medical ethics that would influence both the Islamic world and Europe for centuries.

Al-Razi (Rhazes) and Clinical Medicine

One of the greatest physicians of the Islamic Golden Age was **Al-Razi** (*Rhazes*), whose contributions to clinical medicine and pharmacology were groundbreaking. Al-Razi's *Kitab al-Hawi* (*Comprehensive Book of Medicine*) was an extensive medical encyclopedia that synthesized knowledge from Greek, Persian, and Indian traditions while introducing original observations. Al-Razi was a pioneer in the scientific method, emphasizing the importance of experimentation and careful observation. He also wrote extensively on ethics, advocating for the compassionate treatment of patients and cautioning against quackery.

Al-Razi's work on infectious diseases, particularly his differentiation between smallpox and measles, showcased his diagnostic acumen and his understanding of disease transmission—concepts that would not be fully appreciated in Europe until centuries later.

Ibn Sina (Avicenna) and the Canon of Medicine

Ibn Sina (*Avicenna*), often called the "Prince of Physicians," wrote *Al-Qanun fi al-Tibb* (*The Canon of Medicine*), a monumental work that served as a standard medical text in both the Islamic world and Europe until the 17th century. The *Canon* systematically classified diseases, treatments, and medicinal substances, offering a comprehensive framework for medical practice. Ibn Sina also explored the relationship between psychology and physiology, emphasizing the importance of mental health in overall well-being.

Al-Zahrawi (Abulcasis) and Surgery

Al-Zahrawi (*Abulcasis*), known as the father of modern surgery, authored the *Al-Tasrif*, a 30-volume medical encyclopedia that included detailed descriptions of surgical instruments and techniques. Al-Zahrawi introduced procedures such as cauterization and the use of catgut for sutures, innovations that would become staples of surgical practice. His meticulous documentation of operations and his emphasis on patient care established him as a foundational figure in the history of surgery.

Geometry and Mathematics: Bridging Abstract and Practical Knowledge

Geometry and mathematics were deeply intertwined in the Islamic Golden Age, reflecting a belief that numerical relationships and spatial forms were manifestations of divine order. Muslim mathematicians made groundbreaking contributions to algebra, trigonometry, and geometric theory, which became essential tools for disciplines such as astronomy, architecture, and engineering.

Al-Khwarizmi and the Birth of Algebra

Muhammad ibn Musa al-Khwarizmi, often regarded as the father of algebra, wrote *Kitab al-Mukhtasar fi Hisab al-Jabr wal-Muqabala* (*The Compendious Book on Calculation by Completion and Balancing*), which introduced systematic methods for solving quadratic and linear equations. Al-Khwarizmi's work not only laid the foundation for modern algebra but also introduced the concept of algorithms (a term derived from his name), which are fundamental to computer science.

Omar Khayyam and Geometric Solutions

Omar Khayyam, best known as a poet, was also a brilliant mathematician. He developed geometric methods for solving cubic equations, combining algebraic and geometric reasoning in innovative ways. Khayyam's mathematical work reflected the era's philosophical emphasis on unifying abstract theory with practical applications.

The Legacy of Islamic Geometry

Islamic geometry also found expression in art and architecture, where intricate patterns reflected mathematical precision and aesthetic harmony. The geometric designs of mosques, such as the Alhambra in Spain and the Great Mosque of Isfahan, exemplify the integration of mathematical knowledge with cultural and spiritual values. These patterns, based on symmetry and repetition, illustrate the belief that geometry was not merely a tool for measurement but a way of contemplating the order of the universe.

The Philosophy of Knowledge in the Islamic Golden Age

Underlying these scientific and mathematical achievements was a profound philosophical commitment to the pursuit of truth. Scholars like **Al-Farabi**, **Ibn Rushd** (*Averroes*), and **Al-Ghazali** grappled with questions about the nature of knowledge, the relationship between reason and faith, and the purpose of human existence. While their views often differed, they shared a belief that science and philosophy were complementary paths to understanding the divine and the natural world.

Ibn al-Haytham (Alhazen) and the Experimental Method

Ibn al-Haytham, often regarded as the father of optics, emphasized the importance of empirical observation and experimentation in scientific inquiry. His work *Kitab al-Manazir* (*Book of Optics*) refuted earlier theories of vision, proposing that light travels in straight lines and that the eye perceives objects by receiving reflected light. Ibn al-Haytham's rigorous methodology anticipated the principles of modern science, demonstrating that reason and observation were not oppositional but interdependent.

Averroes and the Reconciliation of Faith and Reason

Ibn Rushd (*Averroes*), a philosopher and commentator on Aristotle, argued for the compatibility of reason and revelation. In works like *The Incoherence of the Incoherence*, he defended the role of philosophy in Islamic thought, asserting that rational inquiry was essential for understanding the divine. Averroes' ideas influenced both Islamic and European intellectual traditions, shaping debates about the relationship between science, philosophy, and religion.

The Lumières and the Enlightenment Quest for Truth

The French Enlightenment, or the **Lumières**, represents one of the most transformative periods in the history of thought, when philosophy and science became intertwined in an audacious quest to understand existence and improve the human condition. Spanning the late

17th and 18th centuries, the Lumières sought to replace superstition, dogma, and blind tradition with reason, evidence, and critical inquiry. This intellectual movement championed human progress, arguing that the careful application of reason could unlock the mysteries of nature, reform societies, and illuminate the path to a more just and enlightened world. Yet, the Lumières also grappled with profound philosophical questions about the limits of reason, the nature of truth, and the role of science in defining human existence.

One of the central figures of the Lumières was **René Descartes**, whose earlier work laid the foundation for the movement's emphasis on rationalism. Descartes' famous dictum, *Cogito, ergo sum* ("I think, therefore I am"), established the primacy of human reason as the basis for knowledge and existence. While Descartes' ideas were deeply influential, the Lumières expanded upon his framework by integrating empirical science into their vision of human progress. Figures like **Voltaire**, **Denis Diderot**, and **Jean-Jacques Rousseau** built on Cartesian principles while critiquing its abstract nature, emphasizing the need to ground reason in the lived realities of human society and the natural world. This shift from abstract rationalism to applied reason reflected the Lumières' commitment to bridging the gap between philosophical inquiry and practical knowledge.

The Encyclopédie: A Manifesto of Enlightenment

One of the most ambitious projects of the Lumières was the **Encyclopédie**, edited by **Denis Diderot** and **Jean le Rond d'Alembert**, a monumental work that sought to compile and disseminate all human knowledge. More than a mere reference work, the Encyclopédie was a radical manifesto for intellectual freedom, scientific inquiry, and human progress. Its pages encompassed everything from mathematics and natural philosophy to critiques of religion and discussions of social justice, reflecting the Lumières' belief that knowledge was the key to liberating humanity from ignorance and oppression.

The Encyclopédie was grounded in the principle that science and philosophy must serve the public good. For the Lumières, knowledge was not an end in itself but a tool for improving society. By organizing and democratizing access to knowledge, they sought to challenge entrenched power structures, particularly the authority of the Church and monarchy, which had long monopolized education and suppressed dissenting ideas. In this sense, the Encyclopédie was both a scientific and political act, embodying the Enlightenment ideal that truth, once illuminated, could transform the world.

Yet, the Encyclopédie also exposed tensions within the Lumières' vision of existence. While many contributors, such as Diderot, embraced a materialist worldview that saw the universe as governed by natural laws, others, like Rousseau, warned against the dangers of unchecked progress. Rousseau argued that the Enlightenment's emphasis on reason and science, while liberating in some respects, could alienate humans from their natural state and deepen social inequalities. This debate over the role of science in shaping human existence remains relevant today, reflecting the enduring complexity of the Lumières' legacy.

Voltaire and the Critique of Dogma

No figure better embodies the spirit of the Lumières than **Voltaire**, a sharp-tongued writer and philosopher whose works relentlessly critiqued religious dogma, authoritarianism, and intellectual complacency. Voltaire championed a vision of existence rooted in tolerance, freedom, and the pursuit of truth, arguing that human beings could only flourish in a society that allowed for the free exchange of ideas. His famous work *Candide* satirized the optimistic philosophies of his time, particularly the idea that "all is for the best in this best of all possible worlds," espoused by figures like Gottfried Wilhelm Leibniz. Through biting wit and vivid storytelling, Voltaire challenged readers to confront the harsh realities of human suffering and to take responsibility for improving the world rather than passively accepting it.

Voltaire's commitment to reason was matched by his defense of science as a means of understanding the universe. He was an early advocate of Newtonian physics, introducing Isaac Newton's ideas to a French audience through works like *Éléments de la philosophie de Newton*. For Voltaire, Newton's discovery of universal laws demonstrated the power of science to transcend human limitations, revealing the order and harmony of nature. At the same time, Voltaire was acutely aware of the limits of human knowledge, warning against the arrogance of claiming absolute truths. This tension between scientific optimism and epistemological humility is a hallmark of the Lumières, reflecting their nuanced approach to the relationship between science, philosophy, and existence.

Rousseau and the Critique of Progress

While Voltaire celebrated the power of reason and science, **Jean-Jacques Rousseau** offered a more critical perspective, questioning whether progress truly led to human flourishing. In works like *Discourse on the Arts and Sciences* and *The Social Contract*, Rousseau argued that the pursuit of knowledge and technological advancement often came at the expense of moral and social harmony. He believed that civilization, by prioritizing material wealth and intellectual achievements, had corrupted humanity's natural goodness and created new forms of inequality.

Rousseau's philosophy challenges the Enlightenment ideal that reason and science are inherently liberating forces. Instead, he emphasized the importance of emotion, community, and a return to simpler ways of living. For Rousseau, existence was not defined by the accumulation of knowledge but by the quality of human relationships and the pursuit of genuine freedom. This critique of progress resonates with contemporary debates about the unintended consequences of technological innovation, from environmental degradation to the ethical dilemmas posed by artificial intelligence.

Philosophical Implications:
Rousseau's critique highlights the tension within the Lumières between the pursuit of knowledge and the search for meaning. While science and philosophy provide tools for understanding the universe, they do not necessarily address the existential needs of human beings. Rousseau's emphasis on the emotional and social dimensions of existence reminds us that the quest for truth must be balanced with the quest for connection and purpose.

The Legacy of the Lumières: A Dual Quest

The Lumières revolutionized the way humanity approaches the great questions of existence, uniting science and philosophy in a shared quest for truth and progress. Their vision of reason as the path to liberation continues to inspire, while their debates about the limits of knowledge, the nature of progress, and the role of emotion in human life remain deeply relevant.

At the heart of the Lumières lies a recognition of the dual nature of existence: the objective realities revealed by science and the subjective experiences explored by philosophy. Figures like Voltaire, Diderot, and Rousseau remind us that understanding the world requires both empirical inquiry and ethical reflection. By seeking to illuminate the unknown while grappling with the complexities of the human condition, the Lumières laid the foundation for modern thought, challenging us to continue their quest with the same courage and curiosity.

Philosophy and Science: A Unified Quest Across Greek Antiquity, the Islamic Golden Age, and the Lumières

Science and philosophy, though distinct in their methods, have historically been intertwined in their shared quest to understand existence and uncover the nature of truth. Across three pivotal intellectual eras—the **Greek antiquity**, the **Islamic Golden Age**, and the **French Lumières**—philosophy served as the foundation and guide for scientific inquiry. These eras, separated by centuries and cultures, reveal striking similarities in their approach to knowledge, their recognition of the interplay between reason and observation, and their shared belief in the transformative power of understanding. While each period reflected the unique cultural and historical contexts in which it flourished, they all championed the idea that the pursuit of knowledge is a deeply human endeavor that transcends boundaries.

The Philosophical Roots of Greek Science

In ancient Greece, science emerged not as an independent discipline but as a branch of philosophy. Thinkers such as **Aristotle**, **Plato**, and **Democritus** viewed the study of the natural world as integral to understanding larger metaphysical questions about reality, causation, and the divine. Aristotle's *Metaphysics* begins with the assertion that "all men by nature desire to know," reflecting a belief that the pursuit of knowledge is a fundamental human drive. For the Greeks, science was not merely a means of practical utility but a path to achieving eudaimonia, or human flourishing.

Aristotle, in particular, synthesized empirical observation with philosophical reasoning, establishing a framework that influenced scientific inquiry for millennia. His concept of the **four causes**—material, formal, efficient, and final—was an attempt to explain not only how things exist but why they exist. This holistic approach bridged the physical and metaphysical, emphasizing that understanding the natural world requires both empirical study and philosophical reflection. Similarly, **Plato's emphasis on the realm of forms**—eternal, unchanging ideals—provided a theoretical foundation for the Greeks' belief in universal principles governing both nature and mathematics.

Greek science, deeply philosophical in nature, sought to reveal the underlying order of the cosmos. Whether through Pythagoras' notion of mathematical harmony, Democritus' atomism, or Archimedes' application of geometry to physical problems, Greek thinkers shared a conviction that reason and observation could unveil the laws of the universe. This unity of science and philosophy became a hallmark of Greek intellectual life, demonstrating that the two disciplines were not rivals but partners in the search for truth.

The Philosophical Synthesis of the Islamic Golden Age

The Islamic Golden Age expanded upon the Greek tradition, integrating it with Islamic theology and Persian, Indian, and other intellectual traditions. Central to this synthesis was the belief that science and philosophy were complementary tools for understanding both the natural world and the divine. The Qur'an itself encouraged inquiry, with verses urging believers to observe the heavens, reflect on creation, and seek knowledge. This theological foundation provided a powerful impetus for the scientific achievements of the era.

Philosophers such as **Al-Farabi, Ibn Sina (Avicenna)**, and **Ibn Rushd (Averroes)** embodied this synthesis, arguing that reason and revelation were not in conflict but were two paths to the same truth. Avicenna's *Book of Healing* and *The Canon of Medicine* exemplify this integration, blending Aristotelian logic with empirical observation to create a comprehensive framework for understanding both the body and the cosmos. Similarly, Averroes' commentaries on Aristotle sought to reconcile Islamic theology with Greek philosophy, emphasizing that divine truth is accessible through both reason and faith.

A striking feature of the Islamic Golden Age was its commitment to the ethical dimensions of knowledge. Scholars like **Al-Ghazali** explored the moral implications of scientific inquiry, arguing that knowledge should be pursued not for personal gain but for the betterment of humanity and the glorification of God. This ethos mirrored the Greeks' belief in knowledge as a path to virtue, while adding a spiritual dimension that reflected the Islamic worldview.

The Islamic Golden Age demonstrated that science and philosophy thrive in environments that value pluralism and intellectual openness. By embracing and expanding upon the works of the Greeks, the Islamic world preserved a tradition of inquiry that would later inspire the European Renaissance and Enlightenment.

The Lumières: Reason as Liberation

The **French Enlightenment**, or Lumières, emerged in a context of religious dogma and political authoritarianism, where knowledge was often restricted by the Church and state. For the thinkers of the Lumières, such as **Voltaire**, **Diderot**, and **Rousseau**, science and philosophy were tools of liberation—means of challenging entrenched authority, illuminating ignorance, and promoting human progress. Their intellectual revolution was driven by a deep faith in reason, empiricism, and the capacity of humanity to improve itself through knowledge.

Philosophy during the Lumières was deeply practical, emphasizing the application of reason to societal problems. Voltaire's satirical critiques of religious dogma, Diderot's monumental *Encyclopédie*, and Rousseau's reflections on the social contract all sought to bridge theoretical insights with the realities of human existence. Science, in this context, was not an abstract pursuit but a means of achieving tangible progress in health, governance, and education.

While the Lumières championed reason and science, they were also keenly aware of their limitations. Rousseau, for example, critiqued the Enlightenment's emphasis on material progress, warning that it could lead to moral and social decay if not tempered by a commitment to justice and compassion. This tension between optimism and caution reflected a philosophical maturity, acknowledging that knowledge must be guided by ethical principles.

The Lumières, like the Greeks and the Islamic Golden Age, saw science and philosophy as interdependent. Whether through Diderot's celebration of reason, Voltaire's advocacy for tolerance, or Rousseau's call for social equality, the Enlightenment echoed the earlier conviction that understanding the natural world and grappling with existential questions were inseparable pursuits.

The Common Thread: A Unified Quest for Truth

Despite their differences in time, culture, and context, Greek antiquity, the Islamic Golden Age, and the Lumières share a profound commonality: the belief that science and philosophy are united in their quest to uncover the truths of existence. Each era recognized that understanding the universe requires both empirical observation and philosophical reflection, combining the precision of science with the depth of philosophy to explore not only how the world works but why it matters.

In all three traditions, knowledge was seen as a moral and spiritual endeavor, not merely a technical achievement. The Greeks viewed the pursuit of wisdom as a path to virtue; the Islamic scholars saw it as a way to honor the divine; and the Lumières framed it as a means of liberating humanity from ignorance and oppression. This shared ethos underscores the idea that science and philosophy, at their best, are not isolated disciplines but part of a larger human project: to illuminate the unknown, confront the mysteries of existence, and improve the human condition.

Moreover, all three eras valued the interconnectedness of disciplines. For the Greeks, geometry, medicine, and astronomy were as much philosophical inquiries as practical sciences. For the Islamic Golden Age, the integration of theology, logic, and empirical study reflected a holistic view of knowledge. And for the Lumières, the blending of scientific and philosophical ideas in works like Diderot's *Encyclopédie* demonstrated the power of interdisciplinary thinking.

Conclusion of Shared Vision

Across Greek antiquity, the Islamic Golden Age, and the Lumières, the pursuit of knowledge emerges as a universal and timeless endeavor, transcending cultural and historical boundaries. By uniting science and philosophy, these eras remind us that the search for truth is not merely an intellectual exercise but a deeply human aspiration—a quest to understand the cosmos, our place within it, and the principles that guide our lives.

This shared vision continues to inspire modern thought, challenging us to approach knowledge with the same courage, humility, and dedication that defined these transformative periods in history.

Chapter 9: De Dubio et Progressu

(On Doubt and Progress)

Introduction: The Role of Doubt in the Pursuit of Truth

Doubt has always been a double-edged sword in the human quest for knowledge. On one hand, it challenges established beliefs, drives inquiry, and fosters innovation. On the other hand, it can paralyze action, undermine trust, and lead to skepticism so pervasive that it erodes confidence in the possibility of truth itself. In science and philosophy, doubt is not merely a barrier to overcome but a necessary tool—a force that, when harnessed, propels progress. This chapter explores the intricate relationship between doubt and progress, examining how questioning assumptions has led to breakthroughs in understanding while also reflecting on the limits and dangers of unrestrained doubt.

From the **skeptics of ancient Greece** to the **scientific revolutions of modernity**, and from the cautious empiricism of **Islamic scholars** to the critical inquiry of the **French Lumières**, the story of progress is inseparable from the story of doubt. This chapter argues that doubt is not the antithesis of certainty but its precursor—a mechanism for refining knowledge and ensuring that it withstands the tests of reason and evidence.

Philosophical Doubt: The Skeptics and the Seeds of Inquiry

The philosophical roots of doubt can be traced back to the ancient Greek **skeptics**, who questioned the reliability of sensory perception, reason, and even the possibility of certain knowledge. Figures such as **Pyrrho of Elis** and **Sextus Empiricus** argued that human understanding is inherently limited, and that suspending judgment (epoché) is the only rational response to uncertainty. While this radical skepticism may seem paralyzing, it also had a constructive dimension, encouraging philosophers to critically examine their assumptions and adopt a more cautious approach to claims of truth.

Socratic Doubt

Perhaps the most famous exemplar of doubt in philosophy is **Socrates**, who used his method of questioning to expose contradictions in others' beliefs and, ultimately, to reveal their ignorance. Socratic doubt was not an end in itself but a means of progress, clearing the ground for a more stable foundation of knowledge. By acknowledging what he did not know, Socrates exemplified intellectual humility, a virtue that would later become central to both science and philosophy.

Cartesian Doubt

Centuries later, **René Descartes** elevated doubt to a methodological principle in his quest for certainty. By systematically doubting everything he could—his senses, his body, and even the external world—Descartes arrived at his famous conclusion: *Cogito, ergo sum* ("I think, therefore I am"). This foundational certainty, he argued, could serve as the basis for rebuilding knowledge on firm grounds. Cartesian doubt thus illustrates how skepticism can lead not to nihilism but to progress, providing a method for separating truth from illusion.

Doubt and the Scientific Revolution

In the realm of science, doubt has played a critical role in overturning dogma and advancing understanding. The **Scientific Revolution** of the 16th and 17th centuries was fueled by thinkers who dared to question the Aristotelian worldview that had dominated for centuries. Figures such as **Copernicus**, **Galileo**, and **Newton** exemplified the power of doubt, challenging assumptions about the cosmos and redefining humanity's place within it.

Copernicus and the Heliocentric Model

For centuries, the geocentric model of the universe, with Earth at its center, was accepted as an unquestionable truth. **Nicolaus Copernicus** challenged this view, proposing that the Sun, not Earth, was the center of the solar system. Copernicus' heliocentric model faced resistance from both religious authorities and the scientific establishment, yet his willingness to doubt the orthodoxy paved the way for subsequent discoveries.

Galileo and Empirical Doubt

Galileo Galilei, often called the "father of modern science," combined empirical observation with mathematical reasoning to challenge longstanding beliefs. His use of the telescope revealed moons orbiting Jupiter and the rough surface of the Moon, providing evidence that contradicted the Aristotelian view of celestial perfection. Galileo's insistence on questioning authority—even at great personal cost—demonstrated the transformative power of doubt in the pursuit of truth.

Newton's Synthesis

While **Isaac Newton** is often celebrated for his achievements, he also embodied the humility of doubt. Newton recognized the provisional nature of his work, famously writing, *"I do not know what I may appear to the world, but to myself I seem to have been only like a boy playing on the seashore, and diverting myself in now and then finding a smoother pebble or a prettier shell, whilst the great ocean of truth lay all undiscovered before me."* This acknowledgment of the unknown reflects the scientific ethos of doubt as a driver of progress.

Doubt in the Islamic Golden Age

During the Islamic Golden Age, scholars embraced doubt as a tool for advancing knowledge, combining critical inquiry with faith in the intelligibility of the universe. Thinkers like **Al-Farabi**, **Al-Ghazali**, and **Ibn Sina** questioned inherited ideas, testing them against reason and empirical evidence. Their approach was characterized by a balance between skepticism and trust in the harmony of divine creation.

Alhazen and the Experimental Method

One of the most notable figures in this tradition was **Ibn al-Haytham** (*Alhazen*), who pioneered the use of the experimental method in optics. By questioning earlier theories of vision, such as the notion that the eye emits rays to perceive objects, Alhazen developed a new model based on empirical observation and mathematical reasoning. His work exemplifies the productive role of doubt in refining scientific theories and advancing understanding.

Al-Ghazali and Philosophical Skepticism

The philosopher **Al-Ghazali** brought doubt into the realm of metaphysics, challenging the deterministic philosophies of his predecessors and emphasizing the limits of human reason. While some critics have accused Al-Ghazali of undermining rational inquiry, his skepticism also spurred a more nuanced approach to knowledge, encouraging scholars like Averroes to defend the compatibility of reason and revelation.

The Lumières and the Power of Critical Inquiry

In the **French Enlightenment**, doubt became a tool for challenging authority and fostering progress. The **Lumières**celebrated the freedom to question religious dogma, political tyranny, and intellectual complacency, believing that critical inquiry could illuminate the path to a better world.

Voltaire's Critique of Certainty

Voltaire epitomized the Enlightenment spirit of doubt, using satire and sharp wit to expose the follies of absolutism and blind faith. His critiques of religious and philosophical orthodoxy, such as those in *Candide*, emphasized the dangers of unwarranted certainty. For Voltaire, doubt was not a weakness but a strength, a means of freeing the mind from the chains of dogma and prejudice.

Diderot and the Encyclopédie

The **Encyclopédie**, edited by **Denis Diderot**, embodied the Enlightenment's commitment to critical inquiry. By compiling knowledge from diverse fields and challenging traditional authorities, Diderot and his collaborators sought to democratize learning and empower individuals to think for themselves. Doubt, in this context, was not an end in itself but a catalyst for discovery and progress.

The Paradox of Doubt: Progress and Peril

While doubt is essential for progress, it also carries risks. In excessive forms, it can lead to **paralysis**, **relativism**, and the erosion of shared truths. The modern world, with its proliferation of misinformation and conspiracy theories, exemplifies the dangers of doubt unmoored from evidence and reason. This paradox underscores the need for balance: doubt must be tempered by trust in the methods and principles that allow us to distinguish truth from falsehood.

Whether in philosophy, science, or society, doubt serves as both a challenge and an opportunity—a force that disrupts complacency and compels us to question, explore, and grow. As we move forward in the pursuit of knowledge, we must embrace doubt not as an obstacle but as an essential partner in the search for truth.

Doubt in the Minds of Great Scientists: A Catalyst for Transformation

The history of science is replete with moments when doubt—far from being a hindrance—proved to be the very engine of progress. Renowned scientists have often faced uncertainty, skepticism, and even crises of belief as they grappled with the limits of their knowledge. These moments of doubt did not diminish their contributions; rather, they deepened their insights, driving them to question assumptions, test hypotheses, and seek better explanations for the mysteries of nature. This section delves into the role of doubt in the lives of key scientific figures, illustrating how embracing uncertainty has shaped the trajectory of human understanding.

Einstein's Cosmological Constant: A "Blunder" Turned Insight

Albert Einstein, one of the greatest minds in the history of science, was not immune to doubt. Early in his career, as he formulated the general theory of relativity, Einstein confronted an apparent conflict between his equations and the prevailing belief in a static universe. His equations suggested that the universe should either expand or contract, but the scientific consensus at the time held that the cosmos was unchanging. To reconcile this discrepancy, Einstein introduced the **cosmological constant** ($\Lambda\Lambda$), a term designed to counteract gravitational attraction and maintain a static universe.

Decades later, observations by **Edwin Hubble** revealed that the universe was, in fact, expanding, validating the predictions of Einstein's original equations and rendering the cosmological constant unnecessary. Einstein famously referred to this adjustment as his "greatest blunder." Yet this episode exemplifies the productive role of doubt in scientific inquiry. Einstein's willingness to revisit and refine his ideas in light of new evidence underscores the importance of humility and adaptability in the pursuit of truth. Ironically, modern cosmology has revived the cosmological constant as a means of explaining dark energy, demonstrating that even "blunders" born of doubt can lead to profound insights.

Charles Darwin: Wrestling with Religious and Scientific Doubts

The development of **Charles Darwin's** theory of evolution by natural selection was shaped by profound personal and intellectual doubts. As a young man, Darwin initially aspired to become a clergyman, adhering to the religious orthodoxy of his time. However, his experiences during the voyage of the HMS *Beagle*—particularly his observations of geological processes and the diversity of life in the Galápagos Islands—began to challenge his beliefs.

Darwin's doubts crystallized as he grappled with the implications of his theory. The idea that species evolved through natural selection, driven by random variation and survival pressures, contradicted the prevailing view of divine creation. Darwin delayed the publication of *On the Origin of Species* for more than 20 years, fearing the societal and religious backlash his ideas might provoke. His personal writings reveal an internal struggle, as he wrestled with the theological and existential questions raised by his work.

Ultimately, Darwin's doubts drove him to rigorously test his theory, gathering extensive evidence from comparative anatomy, fossil records, and artificial selection in domesticated species. His willingness to question both scientific and religious dogmas enabled him to develop one of the most transformative ideas in the history of biology. Darwin's journey illustrates how doubt can serve as a crucible for intellectual growth, compelling scientists to confront uncomfortable truths in their search for understanding.

Niels Bohr and the Quantum Revolution

The advent of quantum mechanics in the early 20th century introduced a level of uncertainty that challenged the deterministic worldview of classical physics. **Niels Bohr**, one of the pioneers of quantum theory, embraced this uncertainty, recognizing that it was not a limitation of knowledge but a fundamental feature of reality. Bohr's principle of **complementarity**—the idea that particles can exhibit both wave-like and particle-like behavior depending on the context of observation—was born from his willingness to accept the paradoxical nature of quantum phenomena.

Bohr's work often placed him at odds with contemporaries like **Albert Einstein**, who famously remarked, "God does not play dice with the universe," expressing his discomfort with the probabilistic nature of quantum mechanics. Bohr, by contrast, argued that the indeterminacy inherent in quantum systems was not a failure of scientific understanding but a reflection of the limitations of classical concepts when applied to the subatomic realm. His debates with Einstein over the interpretation of quantum mechanics epitomize the productive tension between doubt and conviction in scientific progress. By embracing uncertainty, Bohr helped lay the foundations for a theory that continues to revolutionize physics.

Marie Curie: Doubt in the Face of the Unknown

Marie Curie's groundbreaking work on radioactivity exemplifies the courage required to navigate scientific doubt. When Curie and her husband, Pierre, began investigating the phenomenon of radiation, they entered largely uncharted territory. The prevailing understanding of atomic structure was rudimentary, and the idea that atoms could emit energy spontaneously was deeply counterintuitive. Despite the skepticism of many contemporaries, the Curies persisted, isolating new elements such as polonium and radium and coining the term "radioactivity" to describe their findings.

Marie Curie faced not only scientific uncertainties but also societal doubts about her abilities as a woman in a male-dominated field. Her achievements—two Nobel Prizes in different scientific disciplines—were the result of relentless experimentation and an unshakable belief in the value of questioning established ideas. Curie's legacy demonstrates that doubt, far from being a weakness, can be a source of strength and resilience in the pursuit of discovery.

John Nash: Doubt, Genius, and Mental Struggles

John Nash, the brilliant mathematician known for his work in game theory, experienced profound doubts that extended beyond the realm of mathematics into his personal life. Nash's groundbreaking contributions, including the Nash equilibrium, transformed economics and strategic thinking, earning him the Nobel Prize in 1994. Yet his work often emerged from a solitary struggle with complex problems, driven by a deep skepticism of conventional approaches.

Nash's doubts took a darker turn as he battled schizophrenia, a condition that blurred the line between genius and delusion. His ability to navigate this mental illness while continuing to contribute to mathematics is a testament to the complex relationship between doubt, creativity, and resilience. Nash's story reminds us that doubt is not solely an intellectual phenomenon—it is deeply human, encompassing both the potential for brilliance and the vulnerability of the mind.

Doubt as the Core of Scientific Methodology

The stories of these scientists illustrate a central truth: doubt is not the enemy of knowledge but its catalyst. The scientific method itself is a structured form of doubt, requiring hypotheses to be tested, evidence to be scrutinized, and conclusions to be provisional. By embracing doubt, scientists ensure that their findings are robust, their theories adaptable, and their inquiries open-ended.

From Einstein's reconsideration of the cosmological constant to Darwin's existential struggles, from Bohr's acceptance of quantum uncertainty to Marie Curie's perseverance in the face of skepticism, these figures demonstrate that doubt is not a sign of weakness but a source of strength. It is the force that compels us to look deeper, think harder, and imagine possibilities beyond the limits of current understanding.

Doubt as a Philosophical Lens on Scientific Progress

Doubt, as both a philosophical and practical tool, has been central to the evolution of human understanding. It challenges dogma, fuels inquiry, and keeps humanity's intellectual compass pointed toward progress. Philosophers and thinkers such as **Friedrich Nietzsche**, **Baruch Spinoza**, **Victor Hugo**, **Karl Popper**, and **Ahmed Zewail** have approached doubt not as a barrier but as a profound and necessary element of growth in science and human thought. Their reflections illustrate how questioning the status quo, embracing uncertainty, and remaining vigilant against complacency have driven some of history's most transformative breakthroughs.

Nietzsche: Embracing the Abyss of Uncertainty

For **Friedrich Nietzsche**, doubt was not merely a tool for intellectual exploration but a fundamental condition of existence. Nietzsche's philosophy sought to dismantle inherited certainties—be they religious, moral, or scientific—arguing that humanity must learn to navigate a world stripped of absolute truths. His declaration that "God is dead" was less an attack on religion than a recognition that traditional sources of meaning and certainty had been eroded by modernity, including the rise of scientific thought.

Nietzsche saw doubt as both liberating and terrifying. In a world without absolutes, humans were free to create their own values, but they also faced the daunting responsibility of forging meaning in the face of existential uncertainty. This perspective resonates deeply with scientific progress, where every discovery raises new questions and exposes the limits of current understanding. Nietzsche's concept of the **Übermensch**—the individual who embraces doubt and uses it as a catalyst for self-overcoming—parallels the scientist who refuses to rest on established paradigms, instead pushing into the unknown.

In the context of science, Nietzsche's insights remind us that doubt is not just a methodological tool but an existential challenge. The pursuit of knowledge demands courage, as it often leads to the disintegration of comforting certainties. Yet, as Nietzsche argued, it is precisely in this struggle that humanity finds its greatest potential for growth and creativity.

Spinoza: Rational Doubt and the Unity of Nature

Baruch Spinoza approached doubt as a rational exercise, integral to the pursuit of knowledge and the understanding of nature. Spinoza rejected dualistic views of reality, proposing instead that everything—mind, matter, and God—was part of a single, unified substance governed by immutable laws. For Spinoza, doubt arose from incomplete understanding, and the task of philosophy and science was to replace ignorance with clear and distinct knowledge.

Spinoza's rationalist framework had profound implications for scientific progress. He believed that nature was intelligible, its mysteries accessible through the careful application of reason. Doubt, in Spinoza's view, was not an end in itself but a starting point—a recognition of the gaps in one's knowledge that could be filled through systematic inquiry. His deterministic philosophy, which held that all phenomena were the result of necessary causes, encouraged a scientific approach that sought to uncover the underlying principles governing the universe.

Spinoza's emphasis on the unity of all things also anticipated the interdisciplinary nature of modern science. Whether in physics, biology, or psychology, his belief in the interconnectedness of existence inspires a holistic approach to inquiry. In this sense, Spinoza's philosophy of doubt aligns with the scientific method's commitment to questioning assumptions and seeking a deeper understanding of the natural world.

Victor Hugo: Doubt and the Human Spirit

While **Victor Hugo** is best known as a literary giant, his works reflect a deep engagement with philosophical and scientific ideas, particularly the interplay of doubt and progress. In novels such as *Les Misérables* and *Notre-Dame de Paris*, Hugo explored the tension between the constraints of tradition and the liberating power of knowledge, often framing doubt as a force that challenges oppression and sparks transformation.

For Hugo, doubt was inseparable from the human spirit's quest for truth. He saw science as a moral enterprise, one that demanded not only intellectual rigor but also compassion and humility. In *Les Misérables*, for example, Hugo celebrates the progress of science and technology while warning against their misuse, arguing that true progress must be accompanied by social and ethical awareness. Doubt, in this context, becomes a check against hubris, reminding humanity of its fallibility and the need for wisdom in wielding power.

Hugo's reflections on doubt resonate with contemporary debates about the ethical dimensions of scientific progress. Whether in the development of artificial intelligence, genetic engineering, or climate science, his call for a balance between innovation and moral responsibility remains profoundly relevant. Hugo's vision of doubt as a driver of both intellectual and ethical growth underscores its central role in shaping a more humane and enlightened future.

Karl Popper: Falsifiability and the Provisional Nature of Science

Karl Popper, one of the most influential philosophers of science in the 20th century, placed doubt at the heart of scientific progress. In his seminal work *The Logic of Scientific Discovery*, Popper argued that science advances not by proving theories but by rigorously testing and attempting to falsify them. For Popper, a theory's scientific value lay in its falsifiability—the ability to make predictions that could, in principle, be proven wrong.

Popper's emphasis on falsifiability transformed the role of doubt in science. Rather than being a hindrance, doubt became a methodological necessity, a way of ensuring that scientific theories remain open to refinement and revision. This approach rejected the notion of absolute certainty, framing scientific knowledge as provisional and subject to change in light of new evidence. Popper's philosophy also challenged dogmatism, emphasizing that no theory, no matter how successful, is immune to doubt.

In a world increasingly dominated by complex scientific challenges, Popper's vision of doubt as a safeguard against complacency is more important than ever. Whether in climate modeling, vaccine research, or quantum mechanics, his insistence on rigorous testing and critical inquiry ensures that science remains dynamic, self-correcting, and adaptable to new realities.

Ahmed Zewail: The Doubts of the Innovator

Ahmed Zewail, a Nobel Prize-winning chemist often referred to as the "father of femtochemistry," exemplified the role of doubt in driving scientific innovation. Zewail's groundbreaking work involved measuring chemical reactions on the femtosecond scale (one quadrillionth of a second), a feat that many believed impossible. Yet Zewail's willingness to question assumptions about the limits of measurement enabled him to develop techniques that revolutionized chemistry and molecular biology.

Zewail often spoke about the importance of embracing uncertainty in scientific exploration. He believed that doubt was not a sign of weakness but a source of creativity, encouraging scientists to think beyond conventional boundaries and imagine new possibilities. Zewail also emphasized the ethical dimensions of scientific progress, arguing that innovation must be guided by a commitment to improving human well-being.

Drawing inspiration from both Islamic and Western intellectual traditions, Zewail saw science as a universal endeavor, one that transcends cultural and political divides. His approach reflects a synthesis of the doubt-driven inquiry of figures like Popper with the ethical and holistic perspectives of thinkers like Spinoza and Hugo. Zewail's legacy demonstrates that doubt, when coupled with vision and determination, can lead to breakthroughs that reshape our understanding of the world.

The Shared Wisdom of Doubt

From Nietzsche's existential challenges to Popper's falsifiability criterion, from Spinoza's rational inquiry to Zewail's innovative spirit, these thinkers illustrate the multifaceted role of doubt in scientific progress. Each approached doubt not as an obstacle but as an essential force—a catalyst for questioning assumptions, refining theories, and seeking deeper truths.

Their insights remind us that doubt is not the antithesis of progress but its foundation. By embracing uncertainty, we open ourselves to new possibilities, challenge the limits of our understanding, and forge pathways to discovery. In the words of Victor Hugo, "Doubt is the engine of intelligence; it is doubt that leads us to the truth."

The Culmination of Doubt: Progress, Humility, and the Ever-Unfolding Quest

The journey through doubt, as explored in this work, reveals a profound truth: it is doubt—not certainty—that has been the engine of humanity's most transformative achievements. From the Greek philosophers who dared to question the nature of reality, to the Islamic scholars who wove doubt into their pursuit of divine and scientific knowledge, to the French Lumières who used skepticism to illuminate pathways to progress, doubt emerges not as an impediment but as an integral part of human thought. In the last chapter, we examined how doubt operates not merely as an intellectual tool but as a profound and dynamic force, shaping the work of great scientists and thinkers. Figures like Nietzsche, Spinoza, Popper, Zewail, and Hugo remind us that doubt is not a single act but a continual process—a living, breathing element of inquiry that evolves alongside our understanding of the universe.

Doubt compels us to challenge the established, to question the unquestionable, and to reconsider the foundations upon which we build our knowledge. Yet this process is not without its costs. It exposes the fragility of certainty, the limits of our understanding, and the precarious balance between skepticism and belief. This duality—the ability of doubt to both illuminate and destabilize—makes it a paradoxical but essential force. In scientific progress, doubt acts as a sieve, filtering out the false and refining the true. In philosophy, it expands the boundaries of thought, pushing us to confront uncomfortable truths and embrace complexity. And in human existence, doubt is the tension that propels us forward, urging us to seek meaning in a world that often appears indifferent to our presence.

The Fragility and Power of Knowledge

One of the most enduring lessons from the history of science and philosophy is the fragility of knowledge. Every era has clung to its certainties, only to see them upended by new discoveries and insights. The Greeks believed in a cosmos governed by perfect spheres and immutable laws, yet the works of Copernicus, Galileo, and Newton revealed a dynamic and ever-changing universe. The Islamic Golden Age expanded upon Greek thought, introducing new methodologies and perspectives, yet it too faced the limitations of its time. The Lumières celebrated the power of reason and empiricism, but figures like Rousseau and Voltaire also recognized the dangers of unbridled confidence in progress. Each of these periods reminds us that knowledge is not static but provisional, a series of approximations that inch closer to truth without ever fully attaining it.

This fragility, however, is not a weakness—it is a source of strength. By accepting the impermanence of knowledge, we create space for growth, innovation, and adaptation. Science thrives precisely because it embraces this principle, treating every theory as a hypothesis to be tested and, if necessary, replaced. Philosophy, too, gains its vitality from the recognition that no single framework can capture the entirety of human experience. The interplay of doubt and knowledge creates a dynamic process that drives progress while preserving humility. It is this balance—between certainty and skepticism, discovery and revision—that allows humanity to navigate the complexities of existence.

The Intersection of Doubt and Discovery: Insights from Science and Philosophy

The tension between doubt and certainty has shaped the lives and work of some of history's greatest minds. Figures such as **Albert Einstein, Nikola Tesla, Georges Lemaître, Stephen Hawking, Roger Penrose**, and **Edwin Hubble**exemplify the transformative power of doubt, as each challenged prevailing paradigms and reimagined the boundaries of human understanding. Their work reveals that doubt, far from being a weakness, is the crucible in which revolutionary ideas are forged. At the same time, philosophers like **Albert Camus** and **Jean-Paul Sartre** offer profound insights into the existential dimensions of doubt, highlighting its role in the human struggle for meaning in an uncertain world. Together, these scientific and philosophical perspectives illuminate the dynamic interplay of questioning, creativity, and progress.

Albert Einstein: Doubt as the Source of Relativity

Einstein's life and work epitomize the creative potential of doubt. In 1905, while working as a patent clerk, Einstein began questioning the Newtonian framework that had dominated physics for centuries. Newton's laws, while remarkably successful, assumed the existence of an absolute space and time—concepts that Einstein found increasingly problematic in light of emerging evidence from electromagnetism. His doubts about these assumptions led to the development of **special relativity**, a theory that fundamentally redefined the nature of space, time, and motion.

Einstein's willingness to question even his own theories was equally significant. Decades after introducing the **cosmological constant** to preserve a static universe, Einstein abandoned it in the face of Edwin Hubble's observations of an expanding cosmos. Yet the cosmological constant would later reemerge as a critical component of modern cosmology, exemplifying how doubt can lead to insights that transcend the original context of their discovery. Einstein's humility in the face of uncertainty—his acknowledgment that "we are like a little child entering a huge library"—reminds us that doubt is not a sign of ignorance but a mark of intellectual rigor.

Nikola Tesla: The Visionary and the Outsider

For **Nikola Tesla**, doubt was both a source of innovation and a defining feature of his solitary path as a scientist and inventor. Tesla's refusal to accept conventional wisdom drove him to develop revolutionary technologies, including alternating current (AC), which became the foundation of modern electrical systems. At the same time, his skepticism of established institutions and methods often placed him at odds with the scientific establishment.

Tesla's doubts about the limitations of existing technology inspired ideas that were far ahead of their time, such as wireless power transmission and renewable energy. Yet his inability to reconcile his visionary ambitions with practical realities often left him marginalized and misunderstood. Tesla's life serves as a testament to the dual-edged nature of doubt: while it can inspire groundbreaking ideas, it also demands a delicate balance between questioning and collaboration.

Philosophically, Tesla's relentless pursuit of innovation echoes the existentialist notion of living authentically—of pursuing one's vision despite societal pressures or personal setbacks. His work embodies Sartre's idea that "existence precedes essence," as Tesla forged his identity through the creative act of questioning and building.

Georges Lemaître: Reconciling Science and Faith Through Doubt

Georges Lemaître, a Belgian priest and physicist, provides a striking example of how doubt can bridge the seemingly disparate realms of science and faith. In the early 20th century, Lemaître proposed what would later become known as the **Big Bang theory**—the idea that the universe originated from a "primeval atom" or singularity. At the time, this hypothesis challenged both the scientific consensus of a static universe and theological interpretations of creation.

Lemaître's approach to doubt was deeply philosophical. He saw no conflict between his religious beliefs and his scientific work, arguing that doubt in one domain could illuminate truths in the other. For Lemaître, the pursuit of scientific knowledge was an act of humility and wonder, a way of uncovering the order of a universe that transcended human understanding. His ability to navigate doubt with grace and intellectual rigor serves as a model for how science and philosophy can coexist in the search for meaning.

Stephen Hawking and the Limits of Knowledge

Few scientists have embodied the interplay of doubt and discovery as profoundly as **Stephen Hawking**. Diagnosed with ALS at the age of 21, Hawking's life became a testament to the power of intellectual resilience. His groundbreaking work on black holes and the nature of time was driven by a willingness to question fundamental assumptions about the universe, including the boundaries between classical physics and quantum mechanics.

Hawking's exploration of **singularities**—points where the laws of physics break down—exemplifies the productive role of doubt in science. These regions of uncertainty forced physicists to confront the limitations of existing theories, paving the way for the development of quantum gravity. Hawking's concept of **Hawking radiation**, which demonstrated that black holes can emit energy and gradually evaporate, challenged long-held assumptions and redefined our understanding of the cosmos.

Philosophically, Hawking's work resonates with **Albert Camus' notion of the absurd**—the tension between humanity's desire for order and the universe's apparent indifference. Like Camus' *Myth of Sisyphus*, Hawking's journey illustrates the courage required to confront the unknown, finding meaning not in certainty but in the act of exploration itself.

Roger Penrose: The Mathematics of Mystery

Roger Penrose, a mathematician and physicist, has spent his career navigating the intersection of doubt and certainty in the quest to understand the universe. Penrose's contributions to the study of spacetime, black holes, and consciousness reflect a commitment to questioning both scientific orthodoxy and philosophical assumptions.

Penrose's work on **twistor theory**, an alternative framework for describing spacetime, exemplifies his willingness to challenge the dominant paradigms of physics. His collaboration with Hawking on singularity theorems demonstrated how rigorous mathematics could illuminate the most enigmatic aspects of the universe, from the origins of black holes to the nature of the Big Bang.

Penrose's philosophical reflections on consciousness and artificial intelligence further highlight the role of doubt in his work. He has argued that human consciousness cannot be fully explained by classical computation, suggesting that quantum processes may play a role in the mind's operations. This perspective, while controversial, exemplifies Penrose's belief that doubt is essential for pushing the boundaries of understanding.

Edwin Hubble: The Expanding Universe of Doubt

The work of **Edwin Hubble** transformed our understanding of the cosmos, yet it was rooted in a deep sense of doubt about humanity's place in the universe. Hubble's observations of galaxies revealed that the universe was not static but expanding, a discovery that challenged centuries of scientific and philosophical thought.

Hubble's willingness to follow the evidence, even when it contradicted established beliefs, exemplifies the scientific ethos of questioning and revision. His discovery of the **redshift-distance relationship** not only validated Lemaître's Big Bang theory but also raised profound philosophical questions about the nature of infinity and the origins of existence.

In the context of philosophy, Hubble's work echoes **Sartre's existentialism**, which emphasizes the importance of creating meaning in a universe without inherent purpose. By mapping the vastness of the cosmos, Hubble forced humanity to confront its smallness, challenging us to find significance not in our centrality but in our capacity for wonder and inquiry.

Philosophy Meets Science: The Common Thread of Doubt

The lives and works of these scientists—Einstein, Tesla, Lemaître, Hawking, Penrose, and Hubble—demonstrate that doubt is not merely a stepping stone to knowledge but a profound and ongoing aspect of the human condition. In their willingness to question, to revise, and to imagine, they embody the existentialist insight that meaning is not given but created through the act of inquiry.

Philosophers like Camus and Sartre remind us that doubt is inseparable from freedom. To doubt is to take responsibility for one's understanding, to resist the comfort of certainty, and to confront the ambiguity of existence with courage and curiosity. As Sartre famously declared, "Man is condemned to be free," for with freedom comes the burden of choice and the necessity of doubt. In the realm of science, this freedom manifests as the perpetual reexamination of theories, the embrace of uncertainty, and the pursuit of truths that may always remain just out of reach.

Camus, in his concept of the absurd, offers a final insight: that doubt is not a weakness to be overcome but a reality to be embraced. Like Sisyphus pushing his boulder, the scientist and philosopher find meaning not in achieving certainty but in the ceaseless act of questioning. It is through doubt that we progress, not toward a final destination but toward a deeper, richer engagement with the mysteries of existence.

Modernity, with its unprecedented advances in science, technology, and global connectivity, has ushered humanity into an era of paradox. We possess tools capable of mapping the genetic code, exploring distant galaxies, and modeling complex systems with extraordinary precision. Yet, alongside these achievements, we face profound uncertainties: climate crises, ethical dilemmas in artificial intelligence, existential threats from nuclear technology, and the fragmentation of truth in an age of information overload.

At the heart of this paradox lies the dialectic of doubt and certainty—a tension that drives both progress and vulnerability. Doubt compels us to question assumptions and seek new knowledge, while certainty provides the stability needed to act and build. In this chapter, we explore how this dynamic shapes contemporary life, drawing insights from historical precedents, modern science, and philosophical perspectives to chart a path forward.

The Certainties of Science and the Persistence of Doubt

Modern science has achieved remarkable certainties, from the laws of thermodynamics to the predictive power of quantum mechanics. These principles underpin much of modern technology and have transformed how we interact with the world. For instance, the precision of satellite navigation relies on Einstein's theories of relativity, while advances in biotechnology enable life-saving treatments derived from molecular biology.

Yet, the deeper our understanding becomes, the more questions arise. **Stephen Hawking**, in *A Brief History of Time*, marveled at the apparent contradictions between general relativity and quantum mechanics, two cornerstones of modern physics that remain incompatible in extreme conditions, such as black hole singularities. Similarly, the discovery of dark matter and dark energy—together accounting for over 95% of the universe's mass-energy—has exposed the profound gaps in our knowledge.

This interplay between certainty and doubt is not a weakness of science but its defining strength. The scientific method thrives on provisionality, treating knowledge as a dynamic process rather than a fixed endpoint. **Karl Popper's philosophy of falsifiability** underscores this ethos: no theory, no matter how elegant, is immune to revision in the face of new evidence. As modern science ventures into uncharted territories, it reminds us that the pursuit of truth is both iterative and infinite.

Philosophy and the Search for Stability in a Fragmented World

While science grapples with empirical uncertainties, philosophy addresses the existential and ethical dimensions of doubt. Modern philosophers such as **Martin Heidegger**, **Jean-Paul Sartre**, and **Simone de Beauvoir** have explored how uncertainty shapes the human condition, particularly in an age where traditional frameworks of meaning have been challenged by secularization, relativism, and technological change.

Heidegger's concept of **"being-toward-death"** highlights the role of uncertainty in shaping human existence. He argued that confronting the inevitability of death forces individuals to live authentically, making deliberate choices rather than conforming to societal norms. This existential perspective resonates with the challenges of modernity, where rapid change and global crises compel humanity to navigate uncertainty with purpose and resolve.

Similarly, Sartre's idea that **"man is condemned to be free"** emphasizes the responsibility that comes with doubt. In a universe without inherent meaning, individuals must create their own values through action and commitment. This freedom can be both exhilarating and burdensome, particularly in a world where technological advancements outpace our ethical frameworks. Sartre's philosophy invites us to see doubt not as paralyzing but as a call to engagement—a reminder that uncertainty is the price of freedom and progress.

The Ethical Challenges of Modern Doubt

The transformative potential of doubt is matched by its ethical challenges. In an age where information is abundant but trust is scarce, the line between healthy skepticism and destructive cynicism has blurred. The rise of misinformation, conspiracy theories, and ideological polarization reflects a crisis of doubt untethered from critical thinking and evidence.

One area where this dynamic is particularly evident is in public health. The COVID-19 pandemic demonstrated both the power and the fragility of modern science. While vaccines were developed at unprecedented speed, public resistance fueled by distrust and misinformation undermined their impact. This tension underscores the ethical responsibility of scientists, educators, and policymakers to communicate uncertainty transparently while building trust in the processes of inquiry and discovery.

Philosopher **Hannah Arendt** warned of the dangers of cynicism, arguing that when people lose faith in the possibility of truth, they become vulnerable to manipulation and authoritarianism. Her insights resonate in today's "post-truth" era, where the erosion of shared realities threatens the foundations of democracy and collective action. To navigate this landscape, we must cultivate a form of doubt that is constructive rather than corrosive—one that questions assumptions without abandoning the search for common ground.

Doubt, Creativity, and the Future of Knowledge

Despite its challenges, doubt remains a wellspring of creativity. Many of history's greatest innovations emerged not from certainty but from a willingness to question established paradigms. **Nikola Tesla's vision of wireless energy transmission**, **Alan Turing's exploration of artificial intelligence**, and **Ada Lovelace's insights into computing** were all rooted in doubts about the limits of existing knowledge. These breakthroughs illustrate how uncertainty can inspire imagination, opening new possibilities for understanding and invention.

In the context of modernity, creativity driven by doubt is essential for addressing global challenges. Climate change, for instance, demands not only technical solutions but also a reimagining of economic systems, cultural practices, and ethical priorities. Similarly, advances in artificial intelligence require not just algorithmic optimization but philosophical reflection on the nature of consciousness, autonomy, and responsibility.

The philosopher **Michel Foucault** argued that true creativity often arises from questioning the structures of power and knowledge that shape our world. In this sense, doubt becomes a form of resistance—a refusal to accept the status quo and a catalyst for envisioning alternative futures. By embracing doubt as a creative force, we can harness its transformative potential to confront the uncertainties of the modern era.

Conclusion: Toward a Constructive Dialectic

The dialectic of doubt and certainty is not a problem to be solved but a dynamic to be embraced. Certainty provides the foundation for action, while doubt ensures that this foundation remains flexible and open to revision. Together, they form a feedback loop that drives progress, balancing the need for stability with the imperative for change.

In modernity, this interplay takes on new urgency. As we confront the ethical dilemmas of artificial intelligence, the existential risks of climate change, and the social fragmentation of the digital age, we must learn to navigate uncertainty with both humility and courage. This requires fostering a culture of critical thinking, empathy, and collaboration, where doubt is not dismissed as weakness but celebrated as a source of strength.

As the philosopher **Albert Camus** wrote in *The Myth of Sisyphus*, "One must imagine Sisyphus happy." To live authentically in the face of uncertainty is not to seek final answers but to find meaning in the act of questioning itself. This spirit of inquiry, grounded in doubt yet open to discovery, is the essence of both science and humanity—a reminder that the journey toward understanding is as valuable as the destination.

The Ethical Dimension of Doubt

Beyond its intellectual role, doubt carries profound ethical implications. It demands that we confront not only the limits of our knowledge but also the consequences of our actions. In science, doubt safeguards against dogmatism, ensuring that theories remain open to scrutiny and that evidence is weighed without bias. But it also imposes a moral responsibility: to use knowledge wisely and to question whether our pursuits serve the greater good. Figures like Victor Hugo and Ahmed Zewail understood this dual responsibility, emphasizing that progress must be guided by compassion, humility, and a commitment to justice.

The ethical dimension of doubt extends to our interactions with one another. In a world increasingly polarized by ideological divides, the willingness to doubt—our own beliefs as well as those of others—can foster dialogue, empathy, and understanding. Doubt tempers the arrogance of certainty, reminding us that even our most deeply held convictions are shaped by perspective and context. It invites us to approach others with curiosity rather than judgment, creating the conditions for collective progress.

The Unfinished Journey: A Conclusion Without Certainty

As we conclude this exploration, it becomes clear that doubt is not a problem to be solved but a condition to be embraced. It is the thread that connects the ancient philosophers gazing at the stars, the medieval scholars dissecting the mysteries of the human body, and the modern scientists unraveling the quantum fabric of reality. It is the force that drives us to question, to learn, and to grow—an unending journey that defies final answers and resists absolute truths.

In this sense, the conclusion of this work is not a conclusion at all. It is an invitation to continue the process of doubt, to remain open to new questions, and to recognize that the pursuit of knowledge is as much about the journey as it is about the destination. Friedrich Nietzsche's call to "live dangerously," Spinoza's vision of rational inquiry, Karl Popper's principle of falsifiability, and the courage of countless scientists and thinkers remind us that doubt is not a threat but an opportunity—a chance to expand the boundaries of what we know and to glimpse the infinite possibilities of what we have yet to discover.

In a world facing unprecedented challenges—climate change, technological disruption, social inequality—the lessons of doubt are more relevant than ever. They teach us to question assumptions, to seek evidence, and to approach complexity with humility and resolve. They remind us that progress is not a straight path but a winding road, marked by uncertainty, setbacks, and moments of profound insight. And they inspire us to keep moving forward, not despite our doubts but because of them, knowing that it is in the act of questioning that we come closest to truth.

As the great poet and philosopher **Rainer Maria Rilke** once wrote, "Be patient toward all that is unsolved in your heart and try to love the questions themselves." In embracing doubt, we honor the spirit of inquiry that has defined humanity's greatest achievements and prepare ourselves for the challenges and discoveries yet to come.

Chapter 10: Ad Principia Novae Cogitationis

(Towards the Principles of New Thought)

Introduction: Beyond Certainty and Doubt

As we step into an era of unprecedented change and complexity, the interplay between doubt and knowledge takes on new significance. The challenges we face—climate change, artificial intelligence, geopolitical instability, and the ethical dilemmas of biotechnology—demand not only technical solutions but also a profound rethinking of how we approach knowledge and decision-making. In this chapter, we explore how the lessons of doubt, as articulated by great thinkers and scientists throughout history, can guide us toward a framework for addressing the questions of our time.

Ad Principia Novae Cogitationis—"towards the principles of new thought"—is both a call to action and a recognition that the paradigms of the past, while foundational, are no longer sufficient on their own. The principles of new thought require us to blend scientific rigor with philosophical depth, to embrace uncertainty as a tool rather than a hindrance, and to cultivate humility and courage in equal measure. This chapter offers a roadmap for applying these principles, weaving together the insights of philosophy, science, and human experience to chart a path forward.

Principle 1: Embrace Interdisciplinary Thinking

The most pressing problems of the modern world do not fall neatly within the boundaries of any single discipline. Climate change, for example, is not merely an environmental issue; it is an economic, social, and ethical challenge that requires insights from physics, biology, sociology, and political science. Similarly, the rise of artificial intelligence raises questions that extend beyond computer science, touching on philosophy, psychology, and the nature of consciousness itself.

To navigate such complexity, we must move beyond siloed thinking and adopt an interdisciplinary approach. This requires more than just collaboration between fields; it demands a shift in mindset, where scientists and philosophers see their work as part of a larger, interconnected quest for understanding. The Islamic Golden Age offers a powerful model for this kind of integration. Scholars like **Ibn Sina** and **Al-Farabi** seamlessly blended medicine, astronomy, and philosophy, recognizing that knowledge in one domain could illuminate truths in another. The spirit of the **Lumières**, too, reminds us of the transformative potential of uniting diverse perspectives in the pursuit of progress.

In the 21st century, interdisciplinary thinking must become the norm rather than the exception. Universities, research institutions, and policymakers must foster environments where diverse fields intersect, encouraging curiosity and collaboration across boundaries. Only by bridging the gaps between disciplines can we hope to address the multifaceted challenges of our time.

Principle 2: Cultivate Epistemic Humility

In a world awash with information, the ability to recognize the limits of one's knowledge is more important than ever. **Epistemic humility**—the acknowledgment that our understanding is always partial and provisional—has been a cornerstone of scientific and philosophical progress throughout history. It was the driving force behind **Socrates' declaration of ignorance**, **Popper's principle of falsifiability**, and **Einstein's admission of his own "greatest blunder."**

Yet humility is often at odds with the culture of certainty that pervades modern society. In politics, media, and even academia, there is a tendency to reward confidence over caution, simplicity over nuance. This culture not only undermines the search for truth but also fosters division and mistrust. To counteract this trend, we must reclaim the value of humility, recognizing that doubt is not a sign of weakness but a mark of intellectual integrity.

Cultivating epistemic humility requires both individual and systemic change. On a personal level, it means remaining open to new evidence, being willing to admit error, and resisting the allure of ideological certainty. On a broader scale, it involves creating systems that reward curiosity and adaptability, rather than dogmatism or conformity. By embracing humility, we can build a culture of inquiry that is resilient, inclusive, and better equipped to navigate the uncertainties of the future.

Principle 3: Foster Ethical Inquiry

The rapid pace of technological and scientific advancement has outstripped our ability to grapple with its ethical implications. From the potential misuse of artificial intelligence to the environmental consequences of industrialization, the decisions we make today will shape the lives of future generations. To ensure that progress is aligned with human values, we must place ethics at the center of our inquiry.

Ethical inquiry begins with asking the right questions: What are the potential consequences of this discovery or innovation? Who stands to benefit, and who might be harmed? How can we ensure that progress is equitable and sustainable? These questions cannot be answered through science alone; they require the insights of philosophy, history, and the humanities.

Figures like **Victor Hugo** and **Ahmed Zewail** understood the inseparability of ethics and progress. Hugo's writings remind us that knowledge without compassion can lead to tyranny, while Zewail's work exemplifies the potential of science to serve humanity. The ethical principles that guided their lives offer a blueprint for how we can approach the dilemmas of the modern era, balancing innovation with responsibility.

Principle 4: Reimagine Education as a Catalyst for Thought

Education is the foundation upon which new thought is built. Yet traditional models of education often emphasize rote learning over critical thinking, memorization over inquiry, and specialization over breadth. To foster the principles of new thought, education must evolve into a catalyst for creativity, skepticism, and interdisciplinary exploration.

The **Lumières** understood the transformative power of education. Diderot's *Encyclopédie* sought to democratize knowledge, breaking down barriers to learning and empowering individuals to think for themselves. In a similar vein, the scholars of the Islamic Golden Age emphasized the importance of lifelong learning, creating libraries and academies that were open to diverse perspectives. These traditions remind us that education is not just about transmitting knowledge but about cultivating the ability to question, connect, and create.

Modern education must build on these legacies by prioritizing critical thinking, ethical reasoning, and interdisciplinary collaboration. This means moving away from rigid curricula and standardized tests toward more flexible, student-centered approaches that encourage curiosity and adaptability. By equipping future generations with the tools to navigate uncertainty, we can ensure that they are prepared to carry forward the principles of new thought.

Principle 5: Embrace Doubt as a Force for Progress

At the heart of the principles of new thought lies a paradox: doubt, often seen as a barrier, is in fact the foundation of progress. The great thinkers and scientists we have explored throughout this book—from **Socrates** to **Galileo**, **Darwin** to **Popper**—recognized that questioning assumptions is not an obstacle to knowledge but its starting point. In an age of rapid change and global challenges, embracing doubt is more important than ever.

Doubt compels us to ask the hard questions: What if our assumptions are wrong? What if there are better ways of doing things? What are we not seeing? By remaining open to uncertainty, we can avoid the complacency of unquestioned truths and create space for new ideas to emerge. Yet doubt must be balanced with action. While it is important to question, it is equally important to experiment, to test, and to build.

In this way, doubt becomes not a source of paralysis but a force for renewal—a reminder that progress is not a linear path but a dynamic process of questioning, learning, and adapting.

Conclusion: Toward an Open Horizon

As we look to the future, the principles of new thought challenge us to embrace complexity, to seek connections across disciplines, and to approach uncertainty with both humility and courage. They remind us that progress is not the result of certainty but of curiosity, creativity, and a willingness to question the status quo. From the philosophical skepticism of ancient Greece to the ethical inquiry of the Lumières, from the experimental rigor of the Islamic Golden Age to the breakthroughs of modern science, the story of doubt is the story of humanity's relentless pursuit of understanding.

The horizon remains open, filled with questions yet to be asked and discoveries yet to be made. In stepping toward this horizon, we honor the legacy of those who came before us while laying the groundwork for those who will follow. The principles of new thought offer a guide, not as a rigid doctrine but as an invitation to think deeply, act responsibly, and embrace the ever-unfolding journey of knowledge.

Chapter 11: Scientia et Humanitas—Bridging Knowledge and Humanity

(Science and Humanity)

Introduction: Knowledge in the Service of Humanity

Throughout history, the pursuit of science has been driven by two intertwined impulses: the desire to understand the universe and the aspiration to improve the human condition. Yet, the relationship between science and humanity is complex, marked by both triumphs and tensions. Scientific discoveries have eradicated diseases, expanded lifespans, and transformed economies, but they have also introduced profound ethical dilemmas, exacerbated inequalities, and created tools of destruction. This duality reflects the inherent power of knowledge: it is neither inherently good nor bad but shaped by the intentions and values of those who wield it.

In this chapter, we explore the intersection of science and humanity, focusing on how scientific progress influences—and is influenced by—the broader social, ethical, and philosophical dimensions of existence. By examining historical and contemporary examples, we seek to understand how science can align more closely with the needs and values of humanity, ensuring that progress is both meaningful and just.

The Double-Edged Sword of Scientific Progress

Scientific advancements have revolutionized every aspect of human life, yet they have also raised profound ethical questions. Consider the development of nuclear energy, a discovery that embodies both the promise and peril of scientific knowledge. **Albert Einstein** and **Robert Oppenheimer**, two central figures in the development of atomic theory and the Manhattan Project, exemplify the moral complexity of scientific breakthroughs. While Einstein's theories of relativity laid the groundwork for understanding energy-matter equivalence ($E=mc2E=mc2$), he later expressed deep regret over the use of his work to create nuclear weapons. Oppenheimer, upon witnessing the first nuclear test, famously quoted the Bhagavad Gita: *"Now I am become Death, the destroyer of worlds."*

This duality is not confined to the atomic age. The industrial revolution, driven by innovations in engineering and chemistry, transformed economies and improved standards of living, but it also contributed to environmental degradation and exploited labor. The rise of artificial intelligence (AI) offers another contemporary example, with its potential to revolutionize healthcare, education, and transportation balanced against concerns about privacy, job displacement, and ethical misuse.

Philosophically, these dilemmas reflect the tension between **instrumental reason**—the use of knowledge as a tool for achieving specific ends—and **moral reason**, which asks whether those ends are justifiable. Thinkers like **Immanuel Kant** and **Hannah Arendt** have grappled with this tension, emphasizing the need for ethical reflection alongside technical expertise. For science to serve humanity, it must be guided by values that prioritize the common good over narrow interests.

The Role of Empathy in Scientific Inquiry

Science often prides itself on objectivity, yet the most impactful scientific endeavors are those that integrate empathy into their aims and methods. **Jonas Salk**, the developer of the polio vaccine, famously refused to patent his discovery, declaring that the vaccine "belongs to the people." His decision exemplifies the idea that scientific progress should prioritize human welfare over personal or financial gain. Similarly, efforts to address global health challenges, such as the eradication of smallpox or the development of mRNA vaccines for COVID-19, reflect the potential of science to alleviate suffering when guided by compassion.

Empathy also plays a critical role in shaping the direction of research. The feminist scholar **Donna Haraway**, for example, has argued that the dominant narratives of science often reflect the perspectives of those in power, marginalizing alternative voices and needs. By incorporating diverse perspectives—those of women, indigenous peoples, and other historically excluded groups—science can better address the full spectrum of human experiences.

This integration of empathy and inclusivity requires rethinking the metrics of scientific success. Instead of focusing solely on technical achievements or economic gains, we must ask: How does this discovery improve lives? Whose needs are being met, and whose are being overlooked? By placing humanity at the center of scientific inquiry, we can ensure that progress is not only innovative but also equitable and compassionate.

Philosophy and the Limits of Scientific Knowledge

While science offers powerful tools for understanding the natural world, it cannot answer all questions. Issues of meaning, purpose, and value fall outside the purview of empirical observation, requiring the insights of philosophy and the humanities. This limitation does not diminish the importance of science; rather, it underscores the need for a complementary relationship between scientific and philosophical inquiry.

Martin Heidegger, for instance, warned against the danger of reducing existence to measurable phenomena, arguing that the scientific worldview, while invaluable, risks overlooking the richness of human experience. Similarly, **Albert Camus** and **Jean-Paul Sartre** explored the existential dimensions of human life, asking how individuals can find meaning in a universe that science reveals to be vast, indifferent, and devoid of inherent purpose. Their work reminds us that while science can explain the "how" of existence, it is up to humanity to grapple with the "why."

This philosophical perspective is particularly relevant in the age of artificial intelligence and biotechnology. As machines become increasingly capable of mimicking human cognition, we must ask: What does it mean to be human? What values should guide the development and use of these technologies? And how can we ensure that they serve humanity rather than diminish it? Philosophy offers the tools to address these questions, grounding scientific progress in a broader understanding of human dignity and purpose.

The Collective Nature of Progress

Science is often portrayed as the work of lone geniuses, but in reality, it is a profoundly collaborative endeavor. The development of the COVID-19 vaccine, for example, involved the contributions of countless researchers, public health officials, and community organizers across the globe. Similarly, the discoveries of **Isaac Newton**, **Marie Curie**, and **Rosalind Franklin** were built on the insights of their predecessors and contemporaries, reflecting the cumulative nature of human knowledge.

The collective nature of progress highlights the importance of fostering inclusive and collaborative scientific communities. This requires addressing systemic barriers that exclude marginalized groups, ensuring that the benefits of scientific discovery are shared equitably, and cultivating a culture of openness and mutual respect. As the philosopher **John Dewey** observed, "The progress of science requires a democratic culture," one in which knowledge is accessible and inclusive, and the fruits of discovery are directed toward the well-being of all.

Conclusion: Bridging Knowledge and Humanity

The relationship between science and humanity is not static but dynamic, shaped by the interplay of discovery, ethics, and social context. To ensure that scientific progress serves humanity, we must cultivate a holistic approach that integrates technical innovation with philosophical reflection and empathetic engagement. This means embracing the complexities and contradictions of progress, recognizing that the same tools that cure diseases can also create inequalities, and that the same knowledge that liberates can also oppress.

Ultimately, the true measure of scientific progress lies not in the accumulation of facts or the development of technologies but in its ability to enrich human lives. By grounding science in the values of empathy, inclusivity, and ethical responsibility, we can transform knowledge into a force for good—a bridge that connects humanity's aspirations with the vast potential of the universe. As we look to the future, this integrated vision of science and humanity offers a path forward, ensuring that progress is both innovative and just.

Afterword

"In concluding this journey, I am reminded that the threads of doubt and discovery are not just themes in history but lived experiences for all of us. As you close these pages, my hope is that you, too, will embrace the profound beauty of questioning, and in doing so, contribute to the ever-unfolding narrative of human progress." Nader Haddad

Acknowledgments

As this book reaches its conclusion, it is with deep humility and immense gratitude that I reflect on the towering figures whose ideas have illuminated this journey. Their intellectual courage, boundless curiosity, and unyielding dedication to truth have shaped not only the course of history but also the pages of this work. To discuss their ideas has been both a privilege and an act of reverence—a dialogue across time and space with minds that have dared to question, to doubt, and to dream.

To **Socrates**, whose relentless questioning reminds us that wisdom begins in admitting what we do not know; to **Plato** and **Aristotle**, whose contrasting visions of idealism and empiricism laid the foundations for the Western philosophical tradition; and to **Democritus**, whose vision of an atomic universe anticipated truths far beyond his time.

To the brilliance of the **Islamic Golden Age**: **Avicenna (Ibn Sina)**, whose synthesis of science and metaphysics bridges worlds; **Alhazen (Ibn al-Haytham)**, the pioneer of experimental inquiry; and **Averroes (Ibn Rushd)**, whose defense of reason echoes still. Their works remind us that the pursuit of knowledge transcends borders and faiths, uniting humanity in a shared quest for understanding.

To the luminous minds of the **Lumières**: **Voltaire**, whose sharp wit exposed dogma; **Diderot**, whose Encyclopédie became a beacon of enlightenment; and **Rousseau**, who challenged us to balance reason with emotion and progress with purpose. Their legacy calls us to wield knowledge as a force for liberation and justice.

To the modern era's revolutionaries: **Albert Einstein**, who reimagined the universe through the lens of relativity and humbly acknowledged the mysteries still unsolved; **Stephen Hawking**, who looked into the abyss of black holes and found the radiance of understanding; **Roger Penrose**, whose mathematical visions stretch the boundaries of thought; and **Georges Lemaître**, whose union of science and faith reminds us that inquiry knows no bounds.

To the existentialists: **Friedrich Nietzsche**, who taught us to find strength in the void and to create meaning in a world stripped of absolutes; **Jean-Paul Sartre**, who celebrated the radical freedom of human choice; and **Albert Camus**, whose vision of the absurd teaches us to find dignity in struggle and joy in the act of questioning itself.

To the modern sages: **Karl Popper**, who gave us falsifiability as the keystone of scientific progress; **Victor Hugo**, whose words imbued knowledge with empathy and hope; and **Ahmed Zewail**, whose pioneering work in femtochemistry proves that doubt, when coupled with determination, can unlock realms unseen.

To all of these great thinkers and countless others who have enriched this book and my own understanding, I owe a debt that cannot be repaid. Their words have been my guideposts, their doubts my inspiration, and their courage a reminder that to question is to honor the human spirit.

Lastly, I am proud to stand as a humble participant in the ongoing dialogue of ideas. To explore the works of such luminous minds is to glimpse the infinite horizon of human potential. Their greatness does not diminish our own capacity to wonder; rather, it calls us to carry forward the torch of inquiry and to illuminate new paths for those who will follow.

To each of them, and to you, the reader who dares to question and to dream, I offer my deepest thanks.

Nader Haddad

The Author

About the Author

Nader Haddad is a financier, poet, and thinker whose work bridges the worlds of science, philosophy, and mathematics. Educated at the **Sorbonne University** and the **University of Oxford**, Nader has pursued a life dedicated to inquiry, creativity, and the search for understanding. With a strong foundation in finance, he combines his professional expertise with a deep passion for intellectual exploration, particularly in areas where disciplines intersect.

Nader's current research focuses on the **Nash equilibrium**, delving into the intricate dynamics of decision-making and strategy that influence economics, society, and beyond. Expanding his intellectual horizons, he also completed a **Master of Science in Astrophysics** at **Liverpool John Moores University**, where his fascination with the cosmos deepened his appreciation for the interplay between the physical universe and human thought.

A resident of **Paris**, Nader finds inspiration in the city's vibrant history of art, literature, and intellectual discourse. As a poet, he reflects on the human condition, blending lyrical expression with philosophical inquiry. Through his writing, Nader invites readers to embrace the complexity of existence, to question assumptions, and to seek meaning in the ever-evolving tapestry of knowledge.

Principia Philosophiæ Scientiæ is a testament to his belief in the power of doubt and discovery, an invitation to explore the profound connections that unite science, philosophy, and humanity.